The Cold War Defense
of the United States

The Cold War Defense of the United States

Strategy, Weapon Systems and Operations

John E Bronson

McFarland & Company, Inc., Publishers

Jefferson, North Carolina

LIBRARY OF CONGRESS CATALOGUING-IN-PUBLICATION DATA

Names: Bronson, John E, 1954– author.
Title: The Cold War defense of the United States : strategy,
weapon systems and operations / John E Bronson.
Description: Jefferson, North Carolina : McFarland & Company, Inc.,
Publishers, 2019 | Includes bibliographical references and index.
Identifiers: LCCN 2019011175 | ISBN 9781476677200 (paperback. : acid free paper) ∞
Subjects: LCSH: United States—Defenses. | Strategic weapons systems—United
States—History—20th century. | Cold War.
Classification: LCC UA23 .B7836 2019 | DDC 355/.03307309045—dc23
LC record available at https://lccn.loc.gov/2019011175

BRITISH LIBRARY CATALOGUING DATA ARE AVAILABLE

ISBN (print) 978-1-4766-7720-0
ISBN (ebook) 978-1-4766-3581-1

Front cover: B-52 Stratofortress (U.S. Air Force photograph) and
USS *Enterprise* (U.S. Navy photograph)

Printed in the United States of America

*McFarland & Company, Inc., Publishers
Box 611, Jefferson, North Carolina 28640
www.mcfarlandpub.com*

Table of Contents

Introduction

World War II had been the largest and most costly in human terms of all the wars to date. Although the U.S. homeland had been attacked, the largest tasks of the war had been freeing other parts of the world from enslavement at the hands of the Axis powers. Once this freeing had been accomplished, the nation looked forward to continuing its development as a country, largely independently, in peaceful coexistence with the other nations.

Prior to and at the beginning of World War II the nations of the world provided for warning of an attack by monitoring their coastlines and borders with their nearest neighbors. In the U.S. this monitoring was mainly conducted at coastal forts established to protect the shipping ports. Additionally, and on a broader scale, military officers assigned to the staffs of a nation's embassies around the world observed the general capability and state of preparedness of foreign militaries. An attack, if it should come, could be detected by the near approach to those coastlines and borders by the forces of potential adversaries. The range of the weapons of the day dictated that an attacker had to be close to the territory to be attacked. At the time nothing was going to come out of the heavens or from the deep of the oceans to harm a nation without warning. Developments during World War II changed all that.

In that recent war, the United States had desired to carry the war to other parts of the world in order to preserve the manufacturing and economic bases at home. As a result, the United States had developed and deployed only partially the capacity to detect and defend against the long-range bomber or the ballistic missile during the war. The U.S. had evolved a submarine-hunting capability during the war to defend the long-range oceangoing convoys. Additionally, this provided the means to isolate battlefield regions from submarine attack. This need to find adversary submarines grew dramatically after the submarine acquired its own long-range weapons.

As a result of advances in the technologies of conflict made during World War II, the nations of the world had to develop defense mechanisms against the long-range bomber, the submarine, and the ballistic missile. These three types of weapons in particular altered forever the range at which an adversary could launch an attack. These changes significantly modified the type of detection needed to provide security from such an attack.

Immediately following World War II the U.S. had a decided advantage in its submarine fleet and long-range bombers, and its singular advantage of atomic weapons. These advantages could be seen by the Soviet Union as reason for them to catch up in these areas, so as to maintain parity with the other victorious Allies. Following World War II it was innovations in these weapons that made the possibility of attack from longer range real. Due to the mistrust which was evolving between the western nations and the Soviet Union, defense against possible attack by such weapons became desirable.

The U.S. generals and admirals who had key roles in building up the Allied forces of World War II had almost all started with very little. These generals and admirals had seen great risk in their nations' being at war with a much more prepared Germany, and to a lesser degree Japan, while they planned and began to assemble their forces. It is not surprising that these officers, having seen their nation at perhaps its most vulnerable time, would promote the postwar policies of maintaining constant vigilance and preparedness. This preparedness was the focus particularly as the Soviet Union grew more and more into the principal potential adversary for the remainder of the former Allies. The motivation on the part of these leaders was simple: protect America from the types of attacks which these very officers had to plan for and carry out on German-occupied Europe and Pacific areas occupied by Japan during the war years.

Doing this would take vast resources, the likes of which only the United States could provide and sustain. The number of the various types of equipment required by this undertaking is staggering. Airplanes, ships, submarines, radars, communications, weapons, would all be required in great numbers. The number of people needed to operate and maintain them was great, but even greater was the number needed to man the bases, train, feed, house, and defend the operators and maintainers of the front-line defense equipment.

The United States had long relied on massive ocean distances as a primary defense mechanism. This was possible because there was no true adversary in the Western Hemisphere. During the war years this allowed the country to focus resources on the production for and deployment of offensive forces to defeat the Axis nations more or less on their home territories.

All of this created the early postwar situation in which, as these newer threats were being acquired by the Soviet Union in this period of increasing tension and mistrust, the United States had very little capacity to detect and defend against these threats. Initially:

- Only the manned bomber could threaten the U.S.
- The submarine was primarily a threat to the ocean shipping lanes.
- The ballistic missile did not have the range to threaten the continental U.S.

Later, it was the evolution in all of these weapons which created the threat to the continental American homeland. The bomber evolved to longer range and greater stealth. The submarine achieved longer range, greater stealth through the snorkel, and even greater capabilities in both categories through nuclear propulsion. Finally the submarine became the launching pad for ballistic missiles. The ballistic missile evolved into a longer-range weapon and achieved greater accuracy. All three of these vehicles became carriers of nuclear weapons.

Once this evolution was sufficient, any of these schemes could threaten great destruction on the continental U.S. within a very short time. The great ocean distances relied on by the U.S. as protection from traditional potential adversaries were of lesser value when the main potential adversary became the Soviet Union. The United States' proximity to the Arctic and North Pole regions made attack from over this region the obvious path since the opposite side of these same regions was the territory of the Soviet Union. Additionally, those very oceans that had provided the buffer zone previously now became the potential launching site for the newer weapons.

Detection of approaching threats, initially the manned bomber, followed by the submarine and the long-range missile, began to evolve quickly following World War II for the

U.S. This book covers the evolution of this defense in the period from the end of World War II to approximately 1990. It focuses on the evolution of the ability to detect and respond to an attack on America by aircraft or missiles launched from either land bases or from the sea. This detection relies heavily on electronic sensors which could indicate the presence of a hostile vehicle, potentially the carrier of these long-range weapons, the primary threat in this Cold War.

This is not a story of death and destruction. It is rather a story of preventing more of those. In the one sense it is the story of what America did to protect itself. It is also the story of what a generation did, on the country's behalf.

1

Politics and Motivations

Changes on the World Scene and U.S. Actions

It is difficult at best to talk about the origins of the Cold War without discussing several significant elements of World War II. This is hardly unique in history; for example, it is often said that to understand the origins of World War II you must first understand the result of World War I. The Allied powers of World War II in Europe—the United States, Great Britain, and the Soviet Union—constituted a working alliance to oppose the territorial expansion plans of Nazi Germany and Fascist Italy (the Axis). Although this alliance did not start at the inception of conflict in Europe, it was firmly in place the last two to three years or so of the war. This alliance played a role in determining the political arrangements for the conclusion of the war, and influenced the postwar European situation. For example, it was by agreement among the leaders of these powers that only an unconditional surrender would be accepted from Germany. It was further agreed that Germany would be divided such that each of the four powers, including France, would have an army of occupation in one of four areas for a time after hostilities had ended.

The Atlantic Charter, an agreement between Britain's Winston Churchill and President Franklin D. Roosevelt in the early stages of the war, held that following the hostilities of World War II, those countries that had been overrun by Fascist aggression should be free to determine their own form of future government. Later, this principle was reiterated in the Yalta Agreement, set down by the Allied leaders near war's end. The Yalta Agreement would be considered more binding in that it was signed by the Soviet Union's Joseph Stalin in addition to Churchill and Roosevelt.[1]

Europe had averaged a conflict roughly every ten years for centuries. The relatively new twist in the 20th century was to drag in the upstart countries from the New World, the founders of which had for various reasons wanted to leave the older countries. Part of the Allies' motive in postwar arrangements was to break this pattern of repetitive conflict.

At the start of World War II, overall U.S. military strategy consisted of maintaining relatively small standing military forces in peacetime. This was coupled with sizable mobilization of both the industrial base and the citizenry to establish whatever expeditionary force was needed to play a part in quelling a crisis. Although a return to this posture was begun immediately following hostilities of that war, the actions by the Soviet Union following the end of World War II led the U.S., and other nations, to prepare for aggression by the Soviets. Such aggression would result in the Soviet domination of other countries and so remove those countries as threats to the USSR.

With the end of the hostilities of World War II, the reduction in military capacity of the western countries was rapid. In the U.S., the people quickly turned attention to pursuits that had been at least second priority during the war mobilization years: consumerism. The huge number of returning servicemen likewise sought education and a settled home and family lifestyle that they could not have while overseas.

Two principal lessons U.S. defense planners took from the World War II experience were:

1. Don't be surprised. Pearl Harbor had been a great loss and a shock. It had placed the country at the disadvantage of having to begin a war mobilization from a point of vulnerability.

2. Air defense works. Both England and Germany survived substantial air attacks by their adversaries. Although their industries and cities had sustained considerable damage from the bombing, this bombing did not bring the war to a conclusion. And the defenses inflicted considerable damage on the attackers. The counter example was Japan. Thanks to Japan's relatively scant air defenses, their adversary operated with virtual freedom over their country. This caused one high-ranking U.S. officer to comment that air operations over Japan were safer than over most U.S. training airfields.

Another lesson learned by the U.S. in World War II was that if the opponent already had fielded an ability to attack when you begin to build capability, your initial response will be primarily defensive. It was 1943 before the U.S. had adequate ability to go on the offensive. For example, it was 1943 when the U.S. began the offensive bombing campaign on Germany. After this initiation Germany recalled many fighter aircraft from their own offensive efforts in Russia. These fighter aircraft were required to defend the German fatherland from aerial attack. The difference is defending territory and outposts against attacking units only, versus going offensive to remove some of the enemy's ability to attack. Both methods remove part of the adversary's capability, but after offensive ability exists, the opponent has to devote greater resources to watching for an attack. That attack could come anytime, anyplace an enemy's capabilities will allow. So the idea is to match the adversary in readiness.

As a result, initial steps taken by the U.S. for defense of the continental homeland were centered on detection of and response to an airborne attack.

The broader preparation was a significant departure from the former U.S. position. The concept of a small standing army in the U.S. dates to the post-Revolutionary period. The framers of the Constitution and Bill of Rights reflected the opposition to a sizable military in peacetime. It was widely thought that the primary purpose of an army in peacetime was to monitor the citizens. These ideas had been formed prior to the Revolution, when the king of England stationed British troops in the colonies. In the early Cold War, the strategy of the U.S. became:

1. provide support for the UN
2. maintain a deployment of forces in both Atlantic and Pacific regions
3. maintain Naval and Air forces in preparedness
4. maintain the U.S. advantage in atomic weapons
5. maintain a modest-size Army ground force
6. maintain a large reserve of citizen-soldiers.[2]

The strategy needed to support the requirement to have occupation forces in both Germany and Japan as a result of the agreements ending the conflict.

In the settlement between the Allied powers following the defeat of Nazi Germany, the German territory was divided into four zones of occupation. Each of these zones was to be occupied by one of the powers. An agreement by the Allied leaders made at the Yalta Conference in February 1945, by the leaders of the U.S., Britain, and the Soviet Union, largely established the territorial limits for military occupation by each nation following the war. With this agreement, the Soviets would occupy the northeast area, the British would occupy northwest, and the U.S. would occupy the remaining southern area of Germany. Later, the western zones were further subdivided to create a French Zone.[3] General Dwight D. Eisenhower, commander of all Allied forces advancing across France into Germany, saw no military reason for the forces of Britain and the U.S. to advance beyond this line that had been established by the political leaders. Doing so would then lead to a withdrawal back to that line at the completion of hostilities.

The capital of Berlin, located deep within what was the Soviet zone of occupation, was similarly divided into four sectors, each to be occupied by the forces of one of the victorious nations. Air access to these sectors of Berlin was provided for by a set of air corridors, each running from the area of Berlin to the American, British, and French zones of occupation. Within a short time, the American, British, and French zones were not significantly different from each other as these nations had the objective of restoring Germany to be a productive yet peaceful nation. For example, the three Western zones shared a common currency and there were no restrictions on travel or commerce between the three zones. The Soviet Union failed to agree to proposals by the three Western powers to restore all of Germany. As a result, the Soviet zone of occupation and the Soviet sector of Berlin were administered separately. A communist form of government was established for these areas, which became East Germany.

Following the war, the fundamental disagreements between the Western democracies and the Communist ideologies of the Soviet Union surfaced and began to affect relations. During the war these differences had been secondary to the need to defeat the common enemy. Following World War II, observers in the U.S. established a clear focus on the differences between themselves and the Soviet Union. The Soviet Union's leadership was committed to the achievement of world domination. Perhaps the most fundamental point was that Communist ideology held that Western-style capitalism was the root cause of the many European conflicts over the years. These deeply held beliefs established the principles that each of the four nations applied to their respective areas of occupation within Germany. Within their respective German occupation zones, the two styles of government among the former Allies were promoted for adoption in their occupation region as the correct model to achieve the desired long-term stability.

The USSR had lost 20,000,000 people and countless towns, villages, and structures at the hands of Germany in World War II. And both attacks by Germany on the Soviet Union in the twentieth century had occurred through Poland. This created the need, as seen by the Soviets, for a buffer zone of territory between the Soviet Union and Germany, including Poland. This buffer would shield the Soviet homeland from aggression by Western capitalist nations. The Soviets began to organize the countries essentially conquered, on their way west to Berlin, as obedient satellites to provide this highly desirable buffer zone between the homeland and the Western democracies.

In 1946 former British Prime Minister Winston Churchill made what was to become a well-known speech in Fulton, Missouri.[4] The speech included the memorable and often quoted statement, "From Stettin in the Baltic to Trieste in the Adriatic, an iron curtain has descended across the Continent." The curtain hid from view, and other access, the countries of Albania, Bulgaria, Czechoslovakia, East Germany, Hungary, Poland, and Romania, as well as the Soviet Union. The Soviet Union included the areas of Lithuania, Estonia, and Latvia. The significance of this to the Western powers was that it severely restricted them from observing military preparations behind that curtain. Beginning in 1946, British and American militaries undertook steps to learn what they could with this curtain in place. This effort was largely limited to ground-based listening posts and aerial reconnaissance operating around the periphery of this group of nations. This bloc would later become the Warsaw Pact, of which the Soviet Union was the dominant player. The relative lack of information on the military capabilities of the Soviet Union and its satellites, and more importantly the intentions of those nations, would lead to the Western powers' seeing these hidden countries as perhaps more capable than they really were. The creation of satellite countries out of the nations listed, and the suppression of other options for those peoples, led many to the conclusion that the Soviets were interested in territorial expansion. While the U.S. government worked to clarify the Communist intentions, it was prudent to establish defense of the homeland. As seen in the next chapter, the U.S. initiated an air defense against Soviet attack in 1946.

Truman Doctrine

The Truman Doctrine has its origins in a speech to Congress on March 12, 1947.[5] In it President Harry Truman set down the principle that the U.S. would oppose Communist domination of other nations. Britain and the U.S. intended to apply the concepts of the

Map of Europe with Iron Curtain overlaid.

Atlantic Charter to the nations of Eastern and Central Europe in the postwar arrangements. Implementation of this concept was hampered by the fact that the Soviet Union physically occupied those areas at the conclusion of hostilities.

The occasion of this Truman speech was that Greece had made a request to the U.S. for aid. Previously Great Britain had been assisting both Greece and Turkey in their recovery from effects of the recent war. Prior to this, Great Britain had informed the U.S. that because of its own difficulties recovering from the war, it was unable to continue to provide that support to these Mediterranean countries. The larger context of the speech was that President Truman was asking that the U.S. Congress appropriate a goodly amount of money to assist Greece and Turkey. This financial support was intended to preserve these nations as free and democratic countries.

The following is quoted from President Truman's speech to a joint session of the U.S. Congress on March 12, 1947:

> One of the primary objectives of the foreign policy of the United States is the creation of conditions in which we and other nations will be able to work out a way of life free from coercion. This was a fundamental issue in the war with Germany and Japan. Our victory was won over countries which sought to impose their will, and their way of life, upon other nations.
>
> To ensure the peaceful development of nations, free from coercion, the United States has taken a leading part in establishing the United Nations. The United Nations is designed to make possible lasting freedom and independence for all its members. We shall not realize our objectives, however, unless we are willing to help free peoples to maintain their free institutions and their national integrity against aggressive movements that seek to impose upon them totalitarian regimes. This is not more than a frank recognition that totalitarian regimes imposed on free peoples, by direct or indirect aggression, undermine the foundations of international peace and hence the security of the United States.
>
> The peoples of a number of countries of the world have recently had totalitarian regimes forced upon them against their will. The Government of the United States has made frequent protests against coercion and intimidation, in violation of the Yalta agreement, in Poland, Rumania, and Bulgaria. I must also state that in a number of other countries there have been similar developments.
>
> At the present moment in world history nearly every nation must choose between alternative ways of life. The choice is too often not a free one. One way of life is based upon the will of the majority, and is distinguished by free institutions, representative government, free elections, guarantees of individual liberty, freedom of speech and religion, and freedom from political oppression. The second way of life is based upon the will of a minority forcibly imposed upon the majority. It relies upon terror and oppression, a controlled press and radio, fixed elections, and the suppression of personal freedoms.
>
> I believe that it must be the policy of the United States to support free peoples who are resisting attempted subjugation by armed minorities or by outside pressures. I believe that we must assist free peoples to work out their own destinies in their own way.

Truman believed that American actions in World War II had been a massive investment in establishing peace and stability in the world. He also held that assisting nations in their recovery would aid in that objective. He argued that assistance to Greece and Turkey was a relatively small additional investment that would protect the original investment.

Between February and June of 1948, a series of political changes occurred in Czechoslovakia resulting in a Communist government. In elections of 1946, Communist Party candidates had won 38 percent of the vote. This election had resulted in several powerful ministers of the government being Communist. As these ministers consolidated their powers, the non-communist ministers objected and subsequently resigned. These changes led to the resignation of the non-communist president.[6] This set of events became known as the

Czechoslovakia coup d'état, and was seen by the Western nations as strongly influenced by the Soviet Union and a departure by the Soviets from the provisions of the Yalta Agreement.

Nowhere in the world were the differences between the communist and democratic political systems more evident than in Berlin. Although the Eastern sector of Berlin was governed according to the Soviet communist model, the four-power agreement for the entire city included unrestricted access to all areas for the forces of all four nations. There was no physical barrier between the democratically governed Western sectors and the communist-governed Eastern sector. As a result, residents of the city freely moved back and forth. This access gave East Berliners a ready view of conditions in the West, and highlighted the generally poorer conditions in the East. These lifestyle differences were evidenced by the quality and availability of food and consumer goods in the Western sectors compared to the East. This awareness by East Berliners, combined with the presence of the American, British, and French military garrisons in West Berlin, constituted an undesirable blemish in the eyes of the Soviet leadership.

In June 1948, Soviet leader Joseph Stalin ordered that all land access to the Allied sectors of Berlin be closed. Rails, roads, and canals were used to transport supplies for Berlin's residents, as well as the Allied garrisons. These transportation paths were shut down in an attempt to starve those Allied garrisons out of existence. The Soviet action was intended to force the Allied garrisons to depart and remove this outpost of Western democracy inside of East Germany. This shutdown of access to the Western sectors is known as the Berlin Blockade.[7]

Although officials of the U.S. government saw that this development made it impractical for the U.S. to remain in Berlin, when the Secretaries of State, Defense, and the Army approached President Truman with the need for a withdrawal, Truman stated "We are going to stay. Period."[8] As a demonstration of opposition to Soviet coercion, Truman ordered U.S. B-29 bombers, the atomic bombers, to be deployed to England and Germany. He instructed the secretaries that the movement of supplies into Berlin would receive the full capacity of the U.S. military cargo-hauling capability.

The response by the three Western powers was to initiate shipments of food and other necessities into West Berlin by the only remaining transportation method: airplanes. This was possible because the air corridors were guaranteed by formal treaty with the Soviets, whereas access by the ground methods were not. For essentially one full year the Allied nations supplied West Berlin with goods delivered by cargo aircraft. Initially planes were flown into the airport in the American sector, Tempelhof, and both an airfield and a water landing area located in the British sector. During the crisis, land located in the French sector was allocated, an airfield built, and flights used this new airport in the city. The runway and parking areas for this new airport, named Tegel, were originally constructed of crushed rubble. This rubble was left from the destruction done to the city by those same Allied armies up to the conclusion of the recent war. This supplying of the more than 2 million people of West Berlin by air will forever be known as the Berlin Airlift.

The Soviets had exploded their first atomic weapon in August of 1949. This altered the prospects of a Soviet attack even with a limited number of aircraft. By November 1948, with the Berlin Airlift well underway, President Truman directed the National Security Council (NSC) to prepare an assessment of the threat posed to the U.S. by the Soviet Union. This assessment was to provide recommendations for actions to be taken by the U.S. in response. This report, NSC 20, characterized the threat to the U.S. as limited; an attack along the northern coastal areas was possible by a limited number of bomber aircraft.[9] However, damage from such an attack could not threaten the life of the country or its ability to oppose a larger Soviet incursion into Western Europe.

By the spring of 1949 it had become clear that the airlift of supplies was able to keep the people of West Berlin alive. In fact the volume of shipments had grown so that the daily ration had been raised from 1250 calories to 1880. Although still low by western standards, this was adequate to allow the West Berliners to resist overtures from the Communists and hold on to continue their democratic existence. The blockade of the transportation routes ended in May 1949. That same month, the three Western zones became West Germany, with the seat of a democratic government located in Bonn.

President Truman ordered another NSC study in January 1950. The report produced by this assignment, NSC 68, is a more complete and far-reaching view of the role of the U.S. in world events.[10] NSC 68 is a foundation stone and road map for the larger set of actions by the U.S. over the next forty years. This report acknowledged the perception that the USSR intended to be the leader of a communist-dominated world. It expressed alarm at the prospect that with greater Soviet domination of additional countries, there could be insufficient might in a coalition of the remaining countries to effectively oppose the Soviet march. The paper clearly asserts that such domination by the USSR would make it difficult for the U.S. to prosper in its goals of peaceful progress in the world of nations, along the lines expressed in the United Nations' charter.

The following is quoted from the NSC 68 report, from the section Background to the Present Crisis.

Within the past thirty-five years the world has experienced two global wars of tremendous violence. It has witnessed two revolutions—the Russian and the Chinese—of extreme scope and intensity. It has also seen the collapse of five empires—the Ottoman, the Austro-Hungarian, German, Italian, and Japanese—and the drastic decline of two major imperial systems, the British and the French. During the span of one generation, the international distribution of power has been fundamentally altered. For several centuries it had proved impossible for any one nation to gain such preponderant strength that a coalition of other nations could not in time face it with greater strength. The international scene was marked by recurring periods of violence and war, but a system of sovereign and independent states was maintained, over which no state was able to achieve hegemony.

Two complex sets of factors have now basically altered this historic distribution of power. First, the defeat of Germany and Japan and the decline of the British and French Empires have interacted with the development of the United States and the Soviet Union in such a way that power increasingly gravitated to these two centers. Second, the Soviet Union, unlike previous aspirants to hegemony, is animated by a new fanatic faith, anti-thetical to our own, and seeks to impose its absolute authority over the rest of the world. Conflict has, therefore, become endemic and is waged, on the part of the Soviet Union, by violent or non-violent methods in accordance with the dictates of expediency. With the development of increasingly terrifying weapons of mass destruction, every individual faces the ever-present possibility of annihilation should the conflict enter the phase of total war.

On the one hand, the people of the world yearn for relief from the anxiety arising from the risk of atomic war. On the other hand, any substantial further extension of the area under the domination of the Kremlin would raise the possibility that no coalition adequate to confront the Kremlin with greater strength could be assembled. It is in this context that this Republic and its citizens in the ascendancy of their strength stand in their deepest peril.

The issues that face us are momentous, involving the fulfillment or destruction not only of this Republic but of civilization itself. They are issues which will not await our deliberations. With conscience and resolution this Government and the people it represents must now take new and fateful decisions.

NSC 68 principally calls for a policy of containment by the United States towards the Soviet Union. This containment was to consist of:

a. Opposing further territorial expansion by the USSR.
b. Assisting western nations to provide for their own defense.
c. Providing the Soviet satellites of Eastern Europe with an alternative to the Soviet domination, i.e., hope.

In June 1950, Communist forces of North Korea attacked South Korea in the first armed conflict since the end of World War II. Occurring in the time frame of the above events in Europe, this attack was considered another communist initiative against a non-communist neighbor. The U.S. through the United Nations responded with force to oppose this attack.

The U.S. assumed an attack by the Soviet Union on Western Europe. Such an attack would require the United Sates to once again oppose the takeover of the European countries. This opposition would include establishing a massive sea lift of men, supplies, and equipment to provide for the expeditionary war to free those European nations. Defense of America, particularly the industrial centers, therefore was a necessary prerequisite to enable the U.S. to have any chance of performing the steps called for in NSC 68.

Threat is defined as a capability to cause harm combined with an intent to cause that harm. The response to the perceived threats by the U.S. was to build large systems of sensors, information management, and defense installations. These preparations allowed detection, tracking, and generation of whatever response was appropriate once having identified a hostile intruder upon the nation. Because a single bomber or ship approaching the nation is not immediately distinguishable from a peaceful counterpart, its existence must be evaluated in combination with other information. This process is often called an intelligence system. This overall process was well learned and evolved during World War II for hostile submarine location and tracking, as well as a similar process for the airplane threat. These systems would be expanded in response to the increasing perception of the Soviet threat in the next decades.

At each of the events in the above sequence, the U.S. in particular and the West in general stepped up preparations against the presumed Soviet attack on free nations.

Detection and Response Forces

Any security system consists of ability to detect the intrusion, combined with a response capability. The simplest illustration is your local police. You and your neighbors are both the potential target and the sensing element of the system, with the police responding to a hostile situation should it occur. Another simple example is the lifeguard. The lifeguard watches the water and responds to any dangerous situation. The same is true with the defense systems discussed in this book. For all systems, it is desirable to detect the hostile intrusion at the earliest possible time to allow the maximum time for response forces to affect the necessary action.

As will be shown in subsequent chapters, much of the response by the U.S. to these world events was initiated during the presidency of Harry Truman. Much more of it came into being during the presidency of Dwight Eisenhower, to be refined and adjusted by the later presidents.

President George Washington said, "To be prepared for war is one of the most effectual means of preserving peace." It is well to review these principles from time to time.

2

Defending Against
Attack by Airplanes

The American experience in World War II defending against air attack was limited to a handful of key industrial areas, primarily because the war was conducted far away from the North American continent. Those key areas which did have defenses were near the east and west coasts, where it was at least possible that the German and Japanese opponents could mount an attack to reach targets in those areas. Vastly more experience was gained in assessing and plotting to overcome the aerial defenses erected by Germany and Japan to prevent the Allies from attacking their cities and factories.

The most common by far of the defensive emplacements by both sides during most of World War II were point defenses with guns. These guns are more technically termed antiaircraft artillery, and the most common types were 90 mm from the U.S. and the widely known 88 mm of German origin. The range of these guns was approximately 11 miles and up to 33,800 feet altitude. To be effective the gun had to be located near the target being defended; otherwise an attacking air element simply had to navigate around any guns located anywhere other than near the intended target, and concentrate on the guns which were located near the target. The targets thus defended were airfields, factories, rail facilities, bridges, port facilities, and associated supporting facilities.

By the end of World War II, the U.S. air defense capability was primarily applied in England defending that country, and the numerous U.S. airfields located there. These defenses were to protect these sites from German air attacks in the form of the V-1 flying bomb, an unmanned jet-propelled aircraft carrying an 1870-lb. warhead. By this time, the point defense scheme for individual sites had given way to lines of defensive installations which an attacking force would have to cross. Among these lines of defense for England were mixed in U.S. units with equipments consisting of the SCR-584 radar, the Type 9 gun director, and the 90 mm M1 gun, illustrated in the next section. These lines of defense were along the south and east coasts of England, facing German-occupied France, Belgium, and Holland. These countries, until liberated by advancing Allied armies, were the launch sites for the flying bombs. Following this liberation, U.S. units were arrayed in Europe proper, protecting the shipping ports vital to the massive supply effort required to keep the multiple armies equipped and moving forward. Antwerp was a noteworthy center of this air defense protection since it was obvious that its port facilities were absolutely essential for the continuance of the Allied advance.

With the advent of the hydrogen or fusion bomb, the point defense tactics employed during and after World War II became obsolete as the newer weapons had sufficient destructive power to damage more than a single target. As a result, tactics to protect a rail center, shipyard, factory complex, airfield, or other relatively confined target had to be replaced with tactics to damage the attacking forces before they could launch weapons. Longer-range area defense was achieved by the location of interceptor aircraft to cover the oceanic approaches. These interceptor aircraft were themselves mobile gun platforms which could range out over hundreds of square miles upon report of unidentified aircraft.

A series of events led to the United States' decision to discontinue the postwar drawdown of military capability and initiate a defense to protect its continental homeland against possible attack by the Soviet Union. The long-range bomber available during the war—the Boeing B-29—became available to the Soviets before war's end because a number of planes had landed in the Soviet Union under emergency conditions. These planes were kept by the Soviets and used to create a nearly identical copy, the Tupolev Tu-4. In 1947 it was learned that the Soviets had a number of these Tupolev Tu-4 long-range bombers. This airplane was able to threaten the U.S. proper, particularly the northeast and northwest regions.

In early 1948, the government of Czechoslovakia came under the control of Communist leaders. Later in 1948 the Soviets cut off Allied ground access to its West Berlin garrison units, causing the Allied response that has been known ever since as the Berlin Airlift. This event essentially terminated any peaceful interaction between the U.S. and the Soviet Union. Then in 1949 the Soviet Union detonated its first atomic weapon. And in 1950 North Korea initiated its attack on South Korea, leading to a multinational response to that situation. These events led directly the U.S. to field air defense units to protect the population and industrial centers of the nation.

To illustrate how the various elements were integrated and to show the capability and sophistication that both sides had reached as a result of evolution during the war, the air defense system for both England and Germany's Fortress Europe are described below.

ENGLAND'S CHAIN HOME SYSTEM

During the 1930s the British developed first the Chain Home (CH) and later the Chain Home Low (CHL), radars for detection of aircraft.[1] The CH radar was positioned along the coast facing Europe and Germany in particular, since these developments were in response to the thickening war clouds over Europe created by German military expansion. Each CH radar station consisted of a set of four fixed towers that radiated the radio frequency (RF) energy and another two towers to receive any reflected energy. The range that an aircraft could be detected depended on altitude. At an altitude of 2000 feet, detection range was 235 miles, at 5000 feet detection range was 50 miles, and at 13,000 feet detection range was 83 miles. These radars could detect an aircraft between the altitudes of 2000 and 15,000 feet. The shortcoming of the CH radar was that it could not detect low, close-in aircraft, so there was a gap in coverage that an enemy could exploit. This created the need for the CHL radar, which had a moveable antenna so as to cover an area by sweeping the direction of the antenna. The CHL radar could detect down to 500 feet initially, and this coverage further evolved to as low as 50 feet.

These radars signals were handled at the "receiver hut" building located adjacent to the antenna towers. Here the cathode ray tube showing the radar pulse and any reflections was interpreted by an operator, and information was plotted on a plotting table. Detections from individual radars were reported by telephone to the Filter Center located at the central fighter operations facility. The filter center's job was to maintain a map of all incoming raiding aircraft, and to filter out any false information such as multiple reports of aircraft received from different radar sites. This filtering operation involved using knowledge of the specific radar's history and known quirks or deficiencies. Each individual radar site could best determine the range from that site to the detected aircraft. It was the filter center that determined the best estimate of the position of the raiders by performing the range-cutting procedure; that is, taking the ranges from two nearby radar stations and determining the intersection point of the arcs representing those ranges. The intersection was often communicated as a point at a certain angle from north, the azimuth, at a given distance from a known location, such as a radar site. The result of the filtering was intended to present only highly reliable information to the operations centers. The filtered radar information was passed on to as many as three operations centers: the command operations center, the group operations center, and a specific sector operations center.

At the operations centers, plotters, positioned around a large map table of the area, had direct telephone connection to the "tellers" who continuously passed on information from the Filter Room. These plotters positioned markers on the map reflecting the information relayed from the radars. The markers were color coded for time of report, so that everyone could tell the age of the information presented by the color of the marker. As time passed and new information came in, the position and color of the markers were updated. If the report for a particular marker did not arrive, the marker for that aircraft or formation remained fixed in position and its color did not match the others on the table, indicating that the information was not as current. Above the map area sat the fighter controller and the gun controller, along with the air battle commander. Tellers would communicate any directions from these officers to air fields, gun emplacements, or a flight commander in the air. All communication paths, telephone or radio, were dedicated to this operation so there was no delay in the transfer of information.

The sector operations center had attached to it a direction-finding (DF) office, which received the direction-finding signals from individual fighter aircraft of that sector in order to plot on its map the location of the defending fighter forces. Each sector had three DF receiving stations to enable determination by cross bearings the locations of friendly aircraft equipped with the required transmitter.

Fighter aircraft were directed to intercept the attacking aircraft by a method which became known as equal angles. The line between the incoming attackers and the fighters to be vectored onto the attackers became the base on an isosceles triangle.

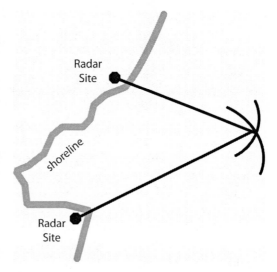

Diagram of two intersecting range cuts from shore-based radar sites.

The intersection of the other two legs of the triangle was the point where the attackers would be intercepted by the fighters. Technically this was true only if the bombers remained on the same course, and the speeds of the two types of aircraft were the same, but this simple method was part of the "good enough" approach the British employed during the defense of their island.

A shortcoming of the system of radar protecting England at this time was that there were no installations in the interior of the country. This meant that after an aircraft crossed the coastline there was no radar information available; the source of information available was from manned observer stations. These observers provided position information based on a simple optical sight and a corresponding calculation of height during clear weather. At night or in cloudy weather these observations were based on the sound of the aircraft.

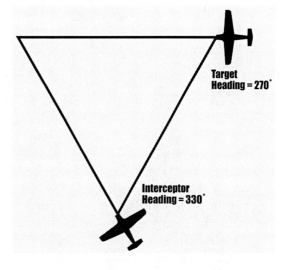

Diagram of the equal angles interception technique.

This system had evolved from 1936 on, through air defense exercises that showed areas for improvement. By 1939 it had reached the state described, which would accurately plot

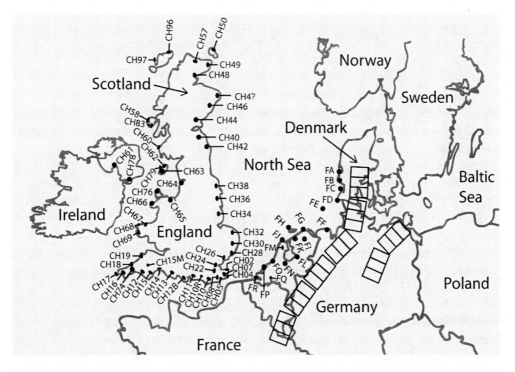

Map of English and German air defenses during World War II. Rectangles indicate the Himmelbett stations of the Kammhuber Line.

the position of unfriendly aircraft, and the information depicted on this plot was no more than four minutes old at any one time. With aircraft speeds of the day, this was adequate.

Germany's Kammhuber Line

Arranged as a similar defensive shield that the Allied air forces, launching from eastern England, had to cross to reach the industrial and transportation centers in Germany was the Kammhuber Line.[2] This line of radar stations, control and communication centers, searchlight installations, gun defenses, and fighter aircraft bases had evolved throughout the war based on the German experiences.

Germany evolved what would be their World War II air defense system considerably more quickly than the British evolution described above. In 1940, because of bombing attacks by British aircraft, Germany started to consolidate its capabilities into a system. There existed the individual capabilities of radar, searchlights, air defense guns known as flak, and fighter aircraft. What was in the favor of the Germans was a more advanced early warning radar in the Freya, and slightly newer still was the Wurzburg radar. The Freya radar had a range of 100 km, or 62 miles, and the Wurzburg 30 km or nearly 19 miles. It was Colonel Joseph Kammhuber, commander of the night fighter forces, who proposed and implemented what the British in particular came to call the Kammhuber Line. This line was made up of boxes called Himmelbetts. These boxes (the English translation of the name means four-poster bed) were each about 20 miles deep in the general direction of toward Berlin, and 27 to 30 miles across generally parallel to the coastline. Each box had a Freya search radar, some collection of searchlights, two Wurzburg tracking radars, and a primary and backup fighter aircraft. Each box also had a control room where the reports were collected and decisions about committing response forces were made by the fighter controller. The two Wurzburg radars were apportioned with one to track the enemy aircraft and one to track the responding fighter aircraft.

The control room in the earlier part of the period was organized like a small theater with bleacher seats. The lower row of seats contained operators who pointed lights at the projection screen opposite the seats: red lights for intercepting fighters and black for attacking bombers. Behind the screen were located other operators who would annotate key information on the screen by reverse writing. Located in the top row of seats were the fighter interceptor controller and communicators to searchlight and antiaircraft artillery units. These centers have become known as "Kammhuber's Cinemas," supposedly by the air crews serving near them. This moniker has probably morphed into Kammhuber Opera Houses in the retelling of the story in the years following the war.

The display screen where the battle was depicted was replaced by the Seeburg Table. The principal element of this table was a large glass top with the map of the area drawn on it, through which lights projected from under the table indicated the location of the attacking and defending aircraft. This table arrangement allowed for the observers around the table to look down at this presentation, similar to that described for the British arrangement previously. A red light indicated the position of the attacking aircraft and a blue light the position of the defending fighter. Another operator marked the progress of the individual lights so that the history and direction of the aircraft could be determined. These track directions allowed the fighter controller to radio direction vectors to the intercepting fighter.

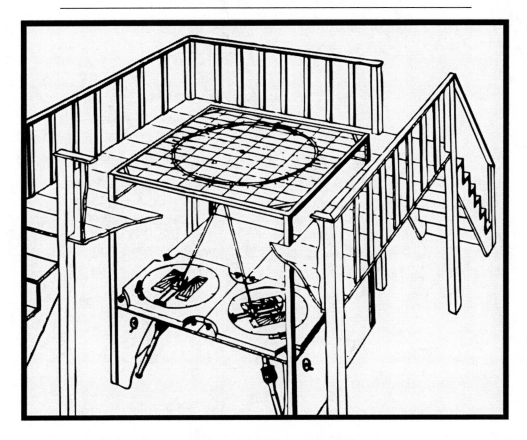

Sketch of a Seeburg Table.

These area air defense centers were located in largely underground buildings constructed of massive amounts of steel and concrete to protect this operation from Allied bombing attacks.

Even though the heaviest bomber attacks on England occurred during the Battle of Britain ending in 1940, the damage done, particularly the loss of life that those attacks caused, motivated the country to maintain and continue to evolve the air defense protection of that island nation. This capability again took center stage in 1944 as Germany fielded the V-1 flying bomb in attacks on England, the London area specifically.[3] This V-1 was launched from a ramp, flew on the course established by the direction of the ramp, and fell to earth when the engine was starved of fuel. In more recent times, this basic description also describes the cruise missile weapon. The air defense establishments of both ground-based guns and aircraft quickly learned to down the subsonic flying weapons.

During the period of the V-1 flying bomb attacks on England, U.S. air defense units comprised part of the total defense. The basic U.S. battery consisted of an SCR-584 Radar, the M9 tracker and director, the power supply, and four 90mm guns. The tracking head element of the equipment contained a pair of telescopes for the two operators to follow the movement of the target aircraft. The operators, the elevation tracker and the azimuth tracker, moved hand wheels to keep the crosshairs of their individual telescopes centered on the aircraft. This movement of the telescope lines-of-sight provided the director with

Sketch of World War II SCR-584 gun battery equipment.

the aircraft's relative motion so that the director could mechanically calculate the aim point for the guns. This aim point was calculated from the present position of the target and used the rate of motion as determined by the tracking telescopes to predict the future position of the target where the gun should place the projectile to intercept the target aircraft. This is called lead computing because the lead angle, from the gun position to the future position of the aircraft, is computed. This firing solution is updated continually while the target is being tracked. The director generates signals to the guns to point them at the target.

The states of the two systems described above, at war's end, were the result of evolution and improvement of the equipment, tactics, and procedures developed over five long years of defending from attacks by the opposing forces. This final state of refinement was the starting point for the evolution of air defenses during the Cold War period.

The speeds of the attacking aircraft of the period immediately following World War II, 350 to 400 MPH, allowed these gun defenses to be adequate. However, two facts related to the speeds of both attacking aircraft and defense weapon projectiles would lead to the need for more capable equipment: the ballistic path followed by the gun-fired projectile had to intercept the flight path of the target aircraft to have any but psychological effect. As the speed of the attacker increased, an earlier and earlier firing time of the gun had to be achieved so that the projectile would be at the predicted point of collision with the attacking aircraft. This gave the gun defense a lower probability of hitting the intruder as the likelihood of the intruder remaining on the predicted line of flight for this length of time became unlikely.

Defending the U.S. from Long-Range Attack

With the advent of the Cold War and the Soviet Union as the principal perceived adversary, the anticipated route of attack was across the Arctic region. This resulted in this region becoming the primary area for detection of hostile aircraft approaching America. The perceived route had to cross much of remote and uninhabited Canada to reach the population and industrial centers in the United States. Such a route across Canada made accessible much of the U.S. heartland. This wide area for potential targets made the ultimate target less predictable, resulting in the need for early detection and tracking. As a result of the concentration of aerospace manufacturing along the coasts during World War II, the U.S. had initiated the establishment of newer factories in the Midwest region, a move somewhat nullified by the perceived threat of bomber attacks which made that region even closer to the launch bases of those aircraft.

As the primary attack weapon became the atomic bomb delivered by aircraft, the con-

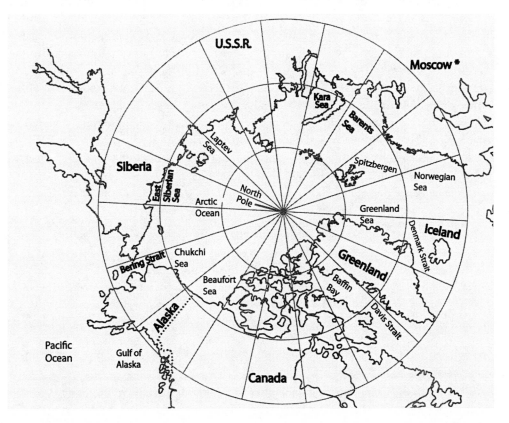

Map of Arctic region north of Canada and the United States, from which an attack by the Soviet Union might come.

cept of a target became an area of targets as compared to the earlier concept of the target as an airfield, bridge, or rail facility. A large city with its accompanying industrial centers, network of transportation facilities, and large concentration of people became the central object to be defended. Additionally, the concept of an attack quickly became that of an all-out attack requiring a full retaliatory response by the U.S. Nearly gone was the concept of attack on one or two specific targets intended perhaps as a stern warning or measured response to some event. The defense concept migrated to that of protecting the nation's ability to respond to a massive atomic/nuclear attack.

Beginning in 1946 the U.S. began to plan for defense against manned bomber attacks against the continental American heartland. These plans initially placed defenses to counter attacks traveling down the flanks of Canada to target areas in the northwest and northeast regions of the U.S. The larger response by the United States was to begin preparations for a continental defense. This meant the initiation of large and long-term projects to provide for detection of aircraft flying from the Soviet Union to target areas in Canada and the United States, and the means to destroy a hostile aircraft that reached these target areas.

The First Radar Networks: Lashup and Permanent

Beginning in 1948, early warning was provided by radars in what was called the Lashup system.[4] These were World War II era radars brought out of storage to provide a perimeter of warning particularly in the northeast and northwest parts of the continental U.S. These sites would report when any unscheduled plane was detected not on a predicted route. Also watching were thousands of civilian volunteers of the Ground Observer Corps. This was organized in 1950 to stand watch and report aircraft movements from 14,000 observation posts reporting to 49 filter centers along the primary approach courses to the key centers of industry and population. The basic training for these observers centered on establishing their ability to identify any U.S. type of aircraft, and hence by dissimilarity, any foreign model. A warning initiated through this network of observers could be passed to any defense center in the country within two minutes.

A radar site of this period consisted of one of the following sets of equipment. Normally both a search radar and a height finder were located on the same site.

Radar	Type	Range	Ceiling	Frequency
AN/CPS-4	Height Finder	90 miles		2700–2900 MHz
AN/CPS-5	Search	60 miles	40,000 feet	
AN/CPS-6	Search	165 miles	45,000 feet	2700–3019 MHz
AN/FPS-3	Search	200 miles		
AN/FPS-4 / TPS-10	Height Finder	200 miles	60,000 feet	9230–9404 MHz
AN/FPS-5	Height Finder			
AN/TPS-1B, 1C	Long Range	120 miles	10,000	1220–1280 MHz

Table of early Cold War search radar capabilities.

The following map depicts the location of U.S. air defense radars in early 1950. Not shown were slightly earlier radars located at Arlington and Hanford, Washington, and at Half Moon Bay, California. The Washington sites provided protection of the Hanford atomic site. Together, these World War II vintage radars also served as test sites for the U.S. to

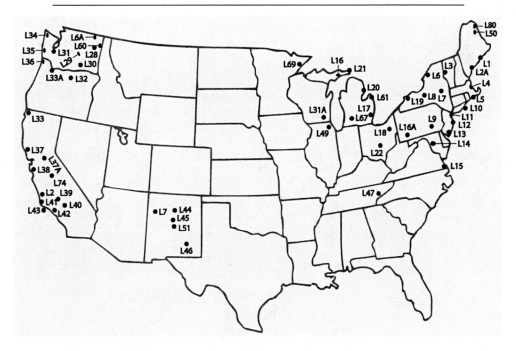

Map of Lashup Radars in the continental United States.

develop and refine its procedures for aircraft interception of manned bomber aircraft. These radar sites performed the dual roles of first detecting the approach of potentially unfriendly aircraft and secondly becoming the sensing basis of the ground-controlled intercept (GCI) operation of dispatching fighter aircraft to investigate and attack. Reports from these radar sites were sent to the Regional Command Center, which in turn reported to the Continental Air Defense command (CONAD) Combat Operations Center at Ent AFB, Colorado. The Regional Command Center dispatched fighter aircraft to investigate the incoming aircraft and to attack if necessary.

Despite the growing public concern about the potential for bomber attacks on the U.S. mainland, the political leadership of the nation resisted proposals by the Air Force, newly created by the 1947 National Security Act, to construct the proposed large network of radars. The initial proposal for a total of 374 radars for the continental U.S. did not go forward due to President Truman's desire to further reduce defense spending following World War II. The proposal for radars was pared down to a list of 75 sites before receiving approval by then Defense Secretary James Forrestal. Even with this approval, funding for these radars was kept out of the government's budget requests until in late 1949 construction started based on funds in the 1950 budget. This initial batch of new construction became known as the Permanent System, as it was to largely replace and expand on the temporary Lashup network.[5]

The following map depicts the location of U.S. air defense radars in mid–1952. These radars were generally known as the Permanent System. Some of the permanent radars were on existing sites from the Lashup system, but with newer equipment. These Permanent radars were constructed in three batches, the first of 24 sites, the second of 28 stations, and the last group of 23, each group waiting for a block of funding to proceed. Over the next several years additional radars were funded as replacements for one of these initial sites,

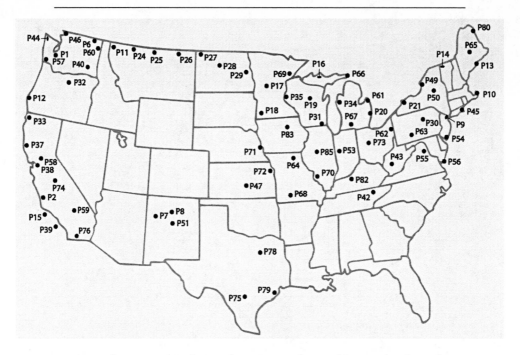

Map of Permanent Radars in the continental United States in mid–1952.

or whenever additional coverage was desired as the exact performance of these planned sites became known through trials and exercises. These additional radars were funded in groups, and although not mobile, a first group was labeled the Mobile Radars and a second group was labeled the Second Mobile Radars.

Each of these radar stations was made up of two parts: the technical site and the domestic site. The technical site was located on a hilltop separate from the domestic site, and primarily housed the radar equipment itself. The domestic site was a small supporting base with offices, living spaces, shops, and some recreational facilities for the people. The staffs were minimal but adequate. Maintenance personnel not needed constantly at one site were generally shared between two sites. For example, Cottonwood, Idaho, was the sister site for Baker, Oregon. The technical site paired with the domestic site at Baker was located on Beaver Mountain, some 13 miles out of town via a two-lane road occasionally blocked by snow. These shared maintenance personnel most often spent three to four days at the other site completing some task. Work at the technical site went on twenty-four hours a day, divided into three shifts of eight hours. Most often, a bus departed the domestic site, with the ongoing crew, sometime before the shift change time, and returned with the crew just off work. As a result of the twenty-four-hour operation of the site, other sections of the domestic site had hours to accommodate all shifts. These stations could be somewhat remote, so the recreational facilities included libraries for readers, perhaps an auto shop, woodworking, and other craft shops to enable the people assigned to make use of their down time.

Similar to the Lashup system, the Permanent radars performed both monitoring of the airspace to detect the unknown aircraft, and then using that same radar to conduct the GCI control of a responding interceptor aircraft. The report from an individual radar site was sent to the Regional Command Center by dedicated phone lines. At the Regional Command Center, the information was used to annotate a plotting board map of the region's area. The Regional Command Center was the central site that accumulated similar information from all the radars in the region, Navy radar picket ships off the coast, and information from the Ground Observer Corps filter centers. This map and information were used by the ground controllers to direct fighter aircraft to intercept any unknown aircraft detected. These commands were passed on to the fighter aircraft units in the region and to the radar site, which would provide the specific guidance to the assigned fighter aircraft dispatched for a particular report. The Regional Command Center also reported both the unknown aircraft detection and the defensive fighter response to the ADC Headquarters at Ent AFB, Colorado, via telephone and teletype. At ADC Headquarters, this information was used to annotate a large map of North America where the information from all the regions was collected. This view of all reported activity concerning the nation could then be monitored and evaluated by the air defense commanders in order to orchestrate the appropriate response at the continental level. The transfer of information by voice and teletype, followed by the manual activity of updating the plotting boards, was time-consuming and prone to errors. This reduced the currency and accuracy of the information available to the command authorities responsible to make timely decisions to adjust the posture and readiness of the defense forces.

Initial Fighter Aircraft Area Defenses

Following detection of approaching aircraft by the radar tripwires, the initial defensive opposition forces consisted of fighter interceptor aircraft. Fighter aircraft of this period carried guns. The fighter would fire the guns from a position behind the target aircraft from a distance of from 200 feet to a maximum of 600 yards. Studies leading up to World War II showed that a little over eight pounds of projectiles was needed to have good probability of bringing down a bomber of the day. This required most fighters to remain in the firing position on a target for two seconds to generate a burst sufficient to deliver this destruction.[6]

In the late 1940s, as the Air Force was planning and initiating the building of an air defense for the continental United States, it also determined that the fighter aircraft available as interceptors were woefully inadequate.[7] The Northrop P-61 Black Widow twin-engine night fighter was old, having been developed during World War II. The North American F-82 Twin Mustang was possibly just not adequate to the task of intercepting an incoming bomber above 25,000 feet. Aircraft under development at that time, the Northrop F-89 Scorpion and the Curtiss F-87, were judged only a marginal improvement. The F-89 was nonetheless slated for production to fill the gap until an all-new fighter could be developed. When the service first took deliveries of the F-89B, then the strengthened F-89C, a number of air crew deaths from crashes due to structural deficiencies in the Scorpion led to lengthy design changes and related aircraft grounding. Rush projects were initiated to provide interim aircraft for defense. Projects to adapt interceptor variants of the Lockheed F-80 and the North American F-86 designs, resulting in the all-weather F-94C and F-86D interceptors, still did not provide interceptors prior to 1953.

The following aircraft provided the manned interceptor response during the period of 1950–1952. This first group of aircraft lacked the de-icing equipment needed to qualify as all-weather.

Aircraft	Range	Speed	Weapons	Fire Control System
F-51	1434 nm	380 knots	six 50 cal.	
F-86E	682 nm	600 knots	six 50 cal., 16 rockets	A-1CM, APG-30
F-89C	2259 nm	550 knots	six 20 mm	E-1
F-94B	700 nm	556 knots	four 50 cal.	E-1
F-9	1175 nm	500 knots	four 20 mm	

Table of early Cold War fighter interceptor capabilities.

The F-51 was the North American P-51 Mustang, one of the principal fighter aircraft of World War II. The F-51 was a single-engine, single-seat pursuit or fighter aircraft, designed and initially delivered during World War II. The definitive version, the P-51D, was powered by the Packard V-1650-7, a license-built version of the Rolls-Royce Merlin 60 series two-stage two-speed supercharged engine. During the Korean War period it was no longer the supreme escort fighter as it had been in the previous war, but its superior qualities led to its adaptation to other roles such as ground attack, reconnaissance, and air defense as discussed here. The F-51 units depicted below were Air National Guard units federalized into national service by the concerns over air attack following the 1948–1950 actions of the Soviet Union. These federalized units were returned to the control of their respective states as additional regular Air Force units came on line near the end of 1952.

The F-86 Sabre, also from North American Aviation, was considered the outstanding fighter aircraft of the 1948–1955 period. The single-seat and single-engine airplane was to become the iconic air-to-air fighter aircraft of the Korean War. Because the initial design would not meet the 600 MPH speed requirement, North American engineers incorporated concepts derived from aerodynamic work in Germany during World War II that had resulted in the Messerschmitt Me 262 near the end of the war. The key concepts so inherited were the 35° swept-back wing and the wing leading edge slats. The six 50-caliber machine guns mounted around the nose fired at 1200 rounds a minute and were normally harmonized to converge 1000 feet in front of the aircraft.

The Northrop F-89 Scorpion was a successor to the largely successful, but after World War II somewhat dated, P-61 night fighter. The new aircraft was powered by two 5,200-lb. thrust Allison J-35 engines mounted under the belly of the slim fuselage, which the two-man crew occupied in tandem seats. The C model mounted six 20 mm forward-firing guns and the crew used the E-1 radar system to position the aircraft in the trail position of the target to initiate firing.

Lockheed's entry into the fighter interceptor arena was in response to an Air Force request in 1949 to make a two-seat interceptor version of the P-80 Shooting Star, America's first operational jet aircraft. This project resulted in the F-94 Starfire, first delivered in the spring of 1950. The F-94 was the first non–German jet night fighter. The F-94 was powered by the 6,000-lb. Allison J33–33 engine with afterburner. The tandem-seated two-man crew used the Hughes E-1 radar to perform the stern-chase attack, which culminated in the firing of the four 50-caliber machine guns mounted in the nose.

The Grumman F-9 Panther was one of the first U.S. Navy carrier-based jet aircraft.

This single-seat, single-engine aircraft was powered by the Pratt & Whitney J48, a license-built version of the British Rolls-Royce Tay engine. Although never assigned to the Air Force's Air Defense Command, these Navy aircraft would have been available in an air defense crisis and are included here to provide the performance comparison.

Other types of aircraft which were assigned to the air defense during this crisis period were the F-47, of P-47 World War II heritage, and the F-84, which was also being used in large numbers in the Korean War.

The following map depicts the locations of U.S. fighter interceptor aircraft in mid–1952.[8] At each installation, four aircraft were kept on alert, two for takeoff in five minutes and two for takeoff in fifteen minutes. All other aircraft were to be able to launch within three hours.

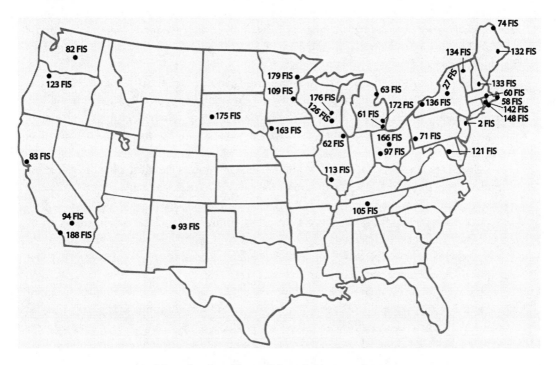

Map of early cold war fighter interceptor bases.

AIR-INTERCEPT PROCEDURE (GUNS)

From prior to World War II, the intercepting fighter aircraft were directed by the ground controlled intercept (GCI) controller, who used the positions of the two aircraft depicted on the radar screen to direct, or vector, the interceptor onto the trail position of an approaching target, from where the attack with guns was made.

The following tactics, techniques, and guidelines were developed and proven during World War II for accomplishing daytime interceptions with conventional, i.e., propeller-driven aircraft. For these cases the radar picture viewed by the controller on the plan position indicator, or PPI, is either at or very near the area being defended. The "day" fighter of this period typically had no onboard radar; the interception was made by visual means when the defending fighter aircraft was in the vicinity of the unknown, or bogey, aircraft. This visual identification would be expected when the two aircraft were within five miles. These methods were taught to all radar controllers, and to all pilots assigned to air defense operations, so that either could be expected to be able to perform these methods whenever needed.

The basic guidelines followed by the GCI controller in directing fighter aircraft to interceptions:

1. Early Vector Imperative—The defending fighters should be sent in the direction of the suspect aircraft as soon as possible after detection. This enables the fighters to intercept the incoming raid at the maximum possible distance from the defended area. This may allow more time for the defenders to operate against the attackers and increases the chances the raid will be intercepted before it can separate into smaller elements, say to attack individual sites within the defended area. If sent too early on an unconfirmed target, the fighters can be recalled.

2. Keep Fighters Joined Up—The incoming raid can be more effectively attacked by the largest number of intercepting fighters that can be assembled for simultaneous attack. It is better to make a larger single attack on the incoming group rather than multiple smaller attacks by separate units. Additionally, this makes the task of following events on the radar scope simpler.

3. Altitude Advantage Imperative—The defending fighters should be positioned at an altitude higher than the highest estimate of the raiding aircraft's altitude. This is a fundamental rule of fighter tactics. The minimum necessary advantage is 2000 feet, but 3000 or 4000 feet advantage is desirable. If the estimate of altitude for the raiding aircraft is unreliable or not available, the defending fighters should be sent to the highest altitude attainable in the time before the expected intercept. The rule of thumb is that no matter how fast a fighter can climb, it can dive 2 or 3 times faster.

4. Stacking for Insurance—When conditions such as when the altitude of the raiding aircraft is not well known, or there are interfering cloud conditions, the fighters may be stacked in altitude. The Flight Leader decides how to apportion the total available force. This should only be done when needed as this makes the radar picture more complex for the controller and divides the defending resources.

5. Backstop for the Break Through—Similarly, when adequate fighter resources are available, a second group of fighters can be positioned approximately ten miles behind the main intercepting force. The backstop would normally have this group of fighters orbit some distance from the point of interception of the raiding aircraft so that any aircraft that evaded the interception force could be intercepted by the backstop fighters.

6. Provide accurate Initial Position Information on Intercepting Fighters—The interception is more easily achieved if the controller has a good estimate of the position of the defending fighters at the time of the initial assignment. This is accomplished by the Flight Leader reporting his position relative to a known geographic point. This allows the controller to make the initial vector assignment more accurately, resulting in a more direct path to the raiding aircraft.

7. Keep Fighters between Raid and Base—This rule leads to keeping the defending fighters on approximately the same bearing as the bogey. The fighter may be vectored slightly

ahead of the bogey bearing to prevent the fighter from entering into a tail chase. When in this position the controller can readily correct the fighter bearing for any course changes observed in the track of the inbound bogey aircraft.

8. Intercept Beyond Visual Range—It is best to make the interception at beyond visual range of the defended area so that the raiding planes do not sight the defended assets. The controller takes into account a number of factors in determining where to affect the intercept:

   a. The need to prevent limited fighter assets being drawn away from the area investigating an unknown.
   b. The need to affect the interception beyond visual and radar range of the likely inbound aircraft.
   c. The desire to affect the interception in an area of good visibility clear of clouds or rain squalls.

9. Advantages of Sun & Clouds—The target aircraft is easier to visually detect when the intercepting fighter is "up-sun." Similarly, the target is easier to spot if silhouetted against a cloud background. The fighter pilot has trained to apply these rules whenever possible. The controller should vector the fighters to take advantage of these concepts if possible; however, the preceding principles should not be sacrificed to achieve these objectives.

10. Keep Flight Leader Fully Informed—Immediately after the Flight Leader reports that he is on the assigned vector it is essential that he is informed of the location of the raid, the estimated altitude, the estimated size of the raid if known, and the heading of the raid. Once the intercepting fighters are within possible visible range of the raid, position of the raiding aircraft should be provided as frequently as possible to aid in the visual sighting.

11. Maintain Radio Discipline—The controller maintains radio discipline by being clear, concise, and quick so that only essential information is transmitted. Once the interception is in progress or combat has commenced, the flight crews have no time for anything but absolutely essential information. In some circumstances, the radio calls can be abbreviated but there can be no possibility of confusion as a result of the shortening of the calls.

Key Communications by the intercepting aircraft:

1. The Steady Report—The Flight Leader reports "Steady" when the intercepting aircraft are on the most recently assigned vector. This allows the controller to compare the heading as flown by the interceptor with the ground track as sensed by the radar, and provide updated heading direction.

2. The Tally Ho Report—When the raiding aircraft are visually identified, it is essential that the pilot making the initial sighting reports the number of the raiding aircraft, their type, and altitude. The "Tally Ho" call indicates he sees them, the report includes: who is reporting the siting, the relative position of the attackers, what is reported (numbers), and at what altitude.

The tracks of unknown aircraft observed by the controller on the PPI scope are divided into two types:

1. Head-On Attack—In this case the unknown aircraft are on a course which will take them directly to or over the defended area, so a response by the defense is required. Sending the

defending fighters out on the same bearing as the bogeys would normally bring the fighters within visual range of the bogeys. A single vector supplied by the controller may be sufficient.

2. Crossing Raid—In this case the unknown aircraft are on a course which does not take them directly to or over the defended area, so this raid may only be monitored by the controller depending on circumstances such as the number of fighters available, and other events nearby. A further course change by the unknowns is required to make this raid a direct threat to the defended area. The defending fighters are provided vectors which place the fighters slightly ahead of the bearing of the bogeys. It is important that the fighters not get behind the bearing of the crossing raid.

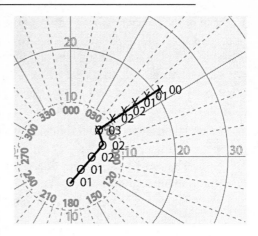

Diagram of Head-On air attack.

The defending aircraft could be utilized in several ways. They could be held at their base until enemy activity was detected which required a response, assigned an orbit point from which they could be dispatched to meet a specific threat, or assigned to patrol in an area to achieve similar readiness.

There are two types of orbit patterns, circular and figure eight. Each pattern can have three different orientations to the track of the aircraft upon entry: centered about the track, left of the track, or right of the track. The figure-

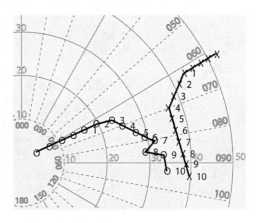

Diagram of Crossing Raid air attack.

eight pattern is generally superior to the simple circular orbit because with the pattern normal to the inbound track to be observed, the pilots have visibility of the inbound track, to the side of the fighter aircraft, for a greater amount of time and they never have their backs turned to that inbound track. The pilots are trained to make tight turns at the ends of the figure eight to make these conditions true. After employing an orbit, if the inbound aircraft are not visually sighted by the time the bogeys are expected to be within five miles of the orbit, the orbiting fighters would be moved closer to the defended area to preclude the possibility of the bogeys getting inside of the defenders. The orbit may be used to position fighters to provide insurance against an incoming raid using the Backstop or Stacking tactics previously described, or for any other reason such as weather or proximity to other action where it is advisable to keep the fighters closer to a potential threat yet within range for other assignments.

There are three types of visual interceptions that may be employed by the controller against unknown aircraft. The selection employed depends on the experience of the controller and on such conditions as number of bogey aircraft, the course, speed, altitude, weather conditions, and the number of fighters available. The previously discussed guidelines apply to all interceptions.

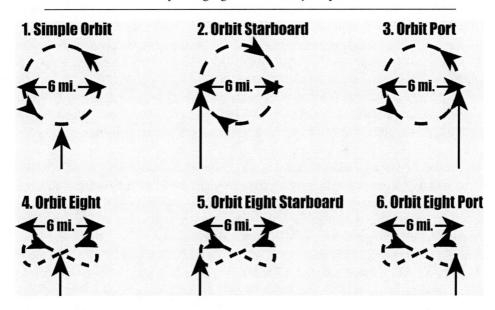

Diagrams of orbit patterns.

1. Head-On Interception—Although called the Head-On to meet a closing aircraft, a more practical method is to offset the fighter aircraft bearing to the target a few degrees so that the visual contact occurs within the 11–1 o'clock. The fighters would normally weave across the intercept vector to aid in this visual search if the target is expected straight ahead. If the Tally Ho does not happen as expected when the bogey is several miles from the fighter, the controller turns the fighter to the approximate course of the bogey to ensure that the fighter remains between the bogey and the defended area.

2. Orbit and Wait Interception—This method is essentially moving the fighters nearer the bogeys so that the inbound track of the bogeys will pass through the orbit pattern of the fighters. In effect the orbit pattern holds the fighters in a defensive position from which they can be directed to a follow-on step towards interception. During the orbit the fighters are performing a standard pattern and the controller is neither directing their flight path nor is he aware of the instantaneous direction of flight. If greater knowledge of the fighters' direction of flight is desired, the standard orbit pattern is replaced by a series of individual heading vectors.

3. Controlled Interception—This method consists of providing a series of heading vectors to bring the fighter aircraft to a position 1 to 3 miles ahead of the bogey, and on a heading similar to the bogey. This method requires both accurate position information on the aircraft involved and unrestricted voice communications with the fighter aircraft.

Examples of interceptions: The following illustrations depict several examples of the intercepts discussed above, in daytime conditions.

These illustrations show the progression of the unknown or bogey aircraft track and the track for intercepting fighter aircraft sent to investigate the intruder. These two tracks of symbols, X's for the intruders and O's for the defenders, and the adjacent notations of the time, would have been drawn on the radar PPI scope surface by the scope operator with a wax pencil.[9]

The ground radar systems of the time would only have been able to show the position of the aircraft at the time of the antenna scan in a particular direction. Phosphor material on the underside of the radar scope surface caused any spot showing reflected radar energy to persist for a time to aid the operator in determining returns of interest. A typical search radar of that time would rotate at 16 revolutions per minute, or 3.75 seconds per revolution. This would provide an update to the scope presentation on the position of any targets at that same rate.

It was these marks made by the scope operator which provided the time history of the motion of the aircraft over the several minutes that the intercept process would take. Based on this annotated history, a determination could be made as to the next step of action that was needed. The scope operator was trained to annotate the PPI presentation at an interval that would facilitate computations of speed, determined by the motion of the dots across the face of the PPI. At each of the times depicted by the marks, the controller communicated to the intercepting aircraft and provided the latest position of the target aircraft based on the most recent radar presentation.

Not depicted in the following illustrations are any other returns visible on the scope such as other aircraft, weather, or any other objects reflecting energy to the radar. The returns of these other objects would complicate the task of the controller to focus on the movements of the two sets of aircraft of interest.

Head-On Interception with opening bogey—The principal events of this interception depicted in the following diagram are:

0800—The track of the inbound is initiated and an initial intercept vector of 30° with an altitude of 10,000 feet at the maximum sustainable speed is assigned to the intercepting fighters.

0801—The fighters report Steady on the assigned vector. The controller is reporting the relative positions of the inbound aircraft as "ahead, 57 [miles]."

0805—The controller now includes in his report of relative position of the inbounds the altitude estimate of 7000 feet.

0809—The controller begins to report the relative positions of the inbound aircraft as "11 o'clock, 12 [miles]" because the distance between the targets and the interceptors has fallen below 20 miles.

0810—In response to the radar plot showing a change in direction of the inbound aircraft, the controller assigns a new vector of 90°to the interceptors. This keeps the heading of the interceptors ahead of the track of the inbound aircraft.

0813—The Flight Leader reports Tally-Ho as "10 o'clock, 2 [miles], 1 aircraft at 6500 feet."

After this visual sighting, the Flight Leader would initiate the engagement with the unknown aircraft.

The following two diagrams show the

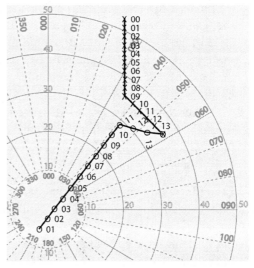

Diagram of Head-On interception with an opening bogey.

sequence for intruders and defenders where an orbit pattern by the defenders is used to maintain the interceptors within some distance of the defended area, yet be in position to attack the intruders if they alter course for the defended area. The left diagram illustrates the case where the sequence results in a visual sighting.

The right diagram illustrates the case where the sequence did not result in a visual sighting by the intercepting aircraft. This indicates that the controller chose not to engage the inbound aircraft. If the sequence continues, the intercepting fighters may be moved to yet another point to wait and see what happens.

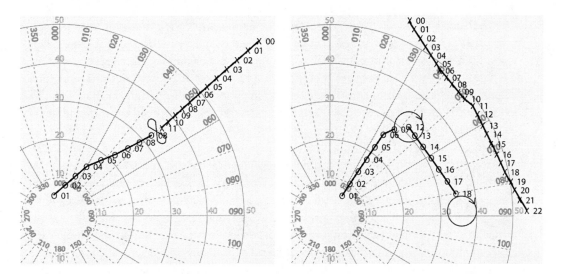

Left: Diagram of Head-On interception utilizing an orbit and wait technique. *Right:* Diagram of interception of a Crossing Raid with multiple orbit and wait sequences.

USE OF THE CONTROLLED INTERCEPTION

The controlled interception provides the controller with the highest degree of control of the intercepting fighters. The controller must have completely up-to-date information on the position of both fighters and bogeys for this to be effective. This essentially requires that the controller is using the radar PPI scope and that it clearly shows all the aircraft involved. The desired outcome of this intercept is for the controller to guide the fighters into position as the flight leader would if he could see the target aircraft. The desired ending position of the intercept is with the fighter essentially on the target's heading, the target is in the 5–7 o'clock position at 1–2 miles range, with the fighter at a higher altitude than the bogey. For a crossing target, the target should be in the 10–11 o'clock position. In all cases this should be accomplished in an area of good visibility so that the flight leader can complete the run on the target.

The key direction provided by the controller is the Safety Vector.[10] This vector brings the fighter aircraft to a course near the track of the target aircraft, but that crosses the track of the target by about 20°. If the fighters and the target are essentially head-on when the safety vector is given, the fighters should be given a vector that is 20–40° beyond the target track. This means a total course change of 200–220° from the interceptor's outbound track.

If the fighters do not report sight of the target aircraft when they cross the track of the target, they may be given a vector to re-cross the track, affecting a zigzag pattern about the radar track of the target. The speeds of the opposing aircraft involved and the turn rates used will determine the distance from the target at which the safety vector is given to the fighter. This 20–40° rule was not applied to a crossing target. If the fighter crossed the track of the bogey, the bogey would be between the fighter and the defended area. Due to the lower closing speed because of the turning fighter, the fighter can approach much closer to the bogey before the safety vector is given. The safety vector in this case would keep the fighter inside the bogey, but on a parallel track within expected visual sighting distance.

Speed and Timing Factors: The interceptor fighter pilots were trained to use standard rate turns. This turn rate is 3° per second, which means completing 360° in 2 minutes. This is commonly called the 2-minute turn. A hard turn would be double that rate, i.e., complete the turn in half the time. The speeds of the two type of aircraft involved in the interception determine the time in advance of the desired interception that the interceptor must be given the safety vector direction. For an interceptor at 280 knots closing head-on with a target at 200 knots, the total distance closing speed is 480 knots, or 8 nautical miles per minute. In order to get on a similar heading as the target, the interceptor will have to complete a turn of nearly 180°, which will consume a minute and 8 nautical miles of the distance between the aircraft. In order to accomplish this, the controller needs to provide the safety vector when the target aircraft is 10 or 11 miles from the interceptor. This added distance provides for any delay by the interceptor in initiating the turn. Such fudge factors are also applied to cover the fact that while in the turn the closing rate between the opposing aircraft is no longer the 8 miles per minute illustrated, and after the turn a new total closing rate will be in effect.

Controlled Interception with offset in bearing—The key events of this sequence are:

0800—The track of the inbound is initiated and an initial intercept vector of 50° with an altitude of 10,000 feet at maximum sustainable speed is assigned to the intercepting fighters.

0801—The fighters report Steady on the assigned vector. The controller is reporting the relative positions of the inbound aircraft as "ahead, 50 [miles]."

0803—The controller now includes in his report of relative position of the inbounds the altitude estimate of 7000 feet.

0806—The controller begins to report the relative positions of the inbound aircraft as "1 o'clock, 18 [miles]" because the distance between the targets and the interceptors has fallen below 20 miles.

0807—The controller directs the interceptors in a turn to the right to a Safety Vector of 200°. This turn and new heading will essentially reverse course and bring the fighters to within about 20° of the track of the inbound aircraft.

0808—The Flight Leader reports Tally-Ho as "8 o'clock, 2 [miles], 1 aircraft at 7000 feet."

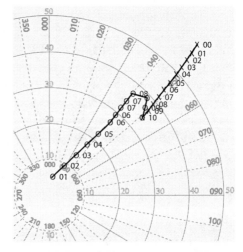

Diagram of a controlled interception with an offset in bearing.

After this visual sighting, the Flight Leader would initiate the engagement with the unknown aircraft.

NIGHT INTERCEPTION

In addition to the above procedures, adjustments and refinements were made to fit the night interception problem, although these procedures are still for conventional piston-propeller powered aircraft. The fundamental difference in the night interception is that the interceptor aircraft is flown essentially on instruments into a position from which the interceptor can make contact with the unknown aircraft with a radar onboard the interceptor. Radar contact provides the interceptor pilot with position information on the target, allowing the interceptor to close on the target. This required a more experienced and focused radar controller, a more capable fighter aircraft, and a fighter pilot with not only special training but attributes, such as patience and confidence in the controller, not found in all fighter pilots.

The night fighter aircraft has the following properties:

a. Onboard radar. The Airborne Intercept (AI) radar provides the range and relative altitude to the bogey aircraft.

b. Heavy firepower.

c. Excellent visibility for the pilot to make identification.

d. Excellent flight stability.

e. Speed.

f. High rate of climb.

g. Ability to decelerate quickly.

h. Endurance.

Night fighters of this period were both single- and multi-engine aircraft. The more common was the larger multi-engine aircraft, which had the advantage of being able to carry more equipment such as radar and weapons, more fuel, and additional crew to operate the radar and gun equipment.

The airborne intercept (AI) radar of this period could search a cone 120° wide to the front of the aircraft. This radar would provide the following selectable modes of operation:

a. Search Mode—this provided the mapping function used for navigation and identification of objects and terrain in front of the aircraft.

b. Intercept Mode—this provided the range, azimuth, and altitude difference from the fighter to the bogey.

c. Gunsight Mode—this provided accurate information allowing the fighter to close within identification and firing range of the bogey.

The search mode had ranges of 65 miles, 25 miles, 5 miles, and 1 mile. The pilot would use these ranges to get nearer the target while centering the selected return for the close approach required for an attack. When the selected target was centered and within one-half mile, the switch was thrown to enter the Gunsight Mode.

The Gunsight Mode, depicted in the diagram below, utilized fixed symbology of a center-segmented circle and two vertical lines etched onto the glass of the display scope. The circle was a reference for the center of the display and the two vertical lines were meaningful after calibration of the radar display. This calibration called for following a friendly

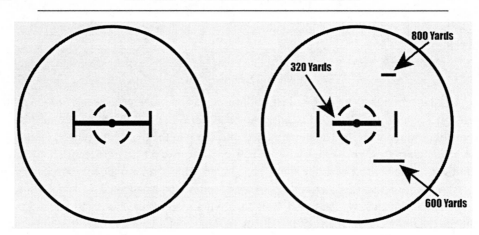

Depiction of World War II airborne radar symbology. Line length indicates target distance.

aircraft at the distance at which the plane's guns were bore-sighted. With the friendly aircraft at this distance, say 250 yards, the calibration control was adjusted so that the width of the returned radar signal just touched the two vertical lines on either sides of the center circle. The left diagram above depicts a target at the same altitude and directly ahead of the radar-carrying aircraft at this calibrated range of 250 yards. The right diagram above depicts what the pilot would see as he closes in on a target from behind. The smallest symbol shown depicts the target at 800 yards ahead, slightly to the right and slightly higher than the interceptor. Another symbol depicts the target at 600 yards ahead, slightly to the right and slightly lower than the interceptor. Another symbol depicts the target at 320 yards directly ahead and at the same altitude as the fighter. Notice that the difference between this depiction and the diagram on the left is only that the return from the target aircraft is larger on the left, showing the closer range. When the pilot had achieved the view depicted on the left diagram on a hostile intruder, he would attack by firing the guns.[11]

For night interceptions, it is absolutely essential that the controller be observing both the defending fighter and the unknown target aircraft on the PPI scope. The controller may be provided information on aircraft speeds and headings that had been extracted from sources such as a plotting table or a dead reckoning trace (DRT) plot, by supporting personnel. The controller needs to provide the interceptor pilot with the simplest instructions possible. Heading should be provided as magnetic, speed and altitude as indicated values, so that the pilot can compare these numbers directly with those seen on the instruments.

The radio communication procedures for the night or low-visibility interception called for more instructions to be delivered more quickly. This is because the night interceptor pilot is highly focused on controlling the aircraft and has only the information provided by the controller until radar contact is established. The pilot must receive the required information as simply as possible, and cannot tolerate other distraction. During the period when the fighter aircraft is being guided into the vicinity of the target aircraft, the following five key pieces of information are transmitted to the pilot, not necessarily in this order.

1. Initial assignment vector which includes the phrase "For Bogey," and the initial heading. This directs the pilot onto the desired heading immediately and indicates that the assignment is an active intercept as opposed to some other activity.

2. Indicated air speed and altitude to approach the target.

3. Position of the bogey.

4. Check on interceptor altitude when fighter is at altitude. This allows the pilot to request a different altitude from what the controller assigned if needed, normally for weather or visibility reasons.

5. Target's heading and airspeed.

Additionally, when moving the fighter to a different heading, the controller would direct the direction of the turn by preceding the heading angle with the direction. For example, "Left two-three-zero." The pilot is kept informed of the target aircraft's position, direction, speed, altitude, and any change observed by the controller's running commentary based on observations of the radar picture. Regardless, the controller communicates not less than once a minute to confirm the communication channel is active.

The intercepting fighter normally had a speed advantage over a typical raiding aircraft, due to the raiding aircraft probably being larger and heavier. This advantage can be used effectively to move the defending fighter closer to the target quickly, and hence further away from the defended area. However, when the fighter is turned onto the trail position of the target, its speed should be reduced to 10 to 30 knots above the target to prevent the fighter from overshooting the target. This is why the fighter used for night interception must be capable of fairly rapid reduction in speed.

As with day interceptions, turns by the interceptor are standard at 3° per second, or the hard turn at 6° per second when needed. The rate of these turns enables the controller to plan the travel of the interceptor during the approach.

The sequence of directions given to the intercepting aircraft during the night interception are the cut-off vector, the safety vector, and the on-course vector. The cut-off vector moves the fighter from its initial location to the vicinity of the target aircraft based on the track history of the target. The safety vector adjusts the track of the fighter to intercept the target's track with approximately a 40° angle from target's track. The on-course vector positions the fighter directly behind the target and on the same heading.

This sequence of directions is embedded within the three main types of night interceptions which the controller may invoke, depending upon the position and direction of the unknown aircraft relative to the defended area. The main types of interceptions are the pure pursuit or curve-of-pursuit, the lead collision or cut-off-vector, and the head-on interception. For all interceptions, the controller is mindful of the amount of time the fighter is in the trail position of the target aircraft, where a rearward firing defensive weapon is usually found. Because the fighter prefers

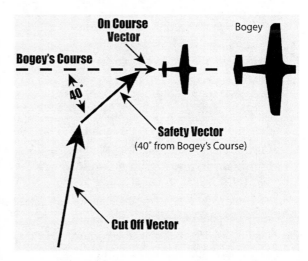

Depiction of air interception terms: Cut Off Vector, Safety Vector, and On Course Vector.

the trail position to best employ gun weapons, the bomber aircraft is often equipped with a weapon specifically designed to protect this area.

Pure Pursuit—The controller provides the pilot with a sequence of heading vectors. Each vector is the heading between the fighter's position and the target's position, so that the fighter is continually pointing at the target. This will bring the fighter in behind the target and will result in the fighter's being on the same heading as the target. The advantage of this approach is its simplicity. The disadvantage of this approach is that it takes longer to accomplish and it exposes the fighter for a longer time in the trail position of the target.

Lead Collision—The controller provides the pilot of the interceptor with a heading vector to a point where the tracks of the two aircraft intersect, bringing the interceptor aircraft to the immediate vicinity of the target aircraft along the shortest possible path. When the fighter is approximately 3 or 4 miles from crossing the track of the target, the controller provides the safety vector, making a 40° angle from target's track. Similarly, when the fighter is ½ to ¾ of a mile from crossing the track of the

![Diagram showing Pure Pursuit with Vector = 280°, Target Heading 270°, and vectors 295°, 330°, 010°, 030°, 040°, 045°, 050° at positions 1-8]

Diagram of sequences of position for an aircraft in Pure Pursuit.

target, the controller provides the on-course vector to place the fighter on the same heading as the target. If the two aircraft have similar speeds, the vector between the fighter's initial position and the initial position of the target, the track of the target, and the collision vector of the fighter are equal in length and form an isosceles triangle. In a previous section this was illustrated as the Equal Angles method and was used by the British air defense system in World War II because of its simplicity.

Head On—The controller provides the pilot of the interceptor with heading vectors to bring the fighter on an opposing course to the target with a 4-mile offset or displacement between the tracks of the two aircraft.

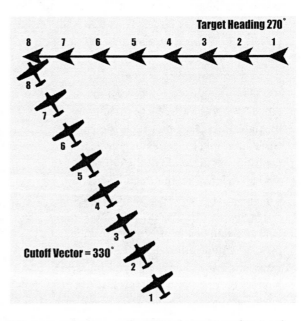

Diagram of sequences of position for an aircraft in Lead Collision.

When the fighter is approximately 5 miles from the target, the controller provides the safety vector, as with the other types of interception, making a 40° angle from the target's track, and then provides the on-course vector. This is a more difficult interception because of the closing speed of the two aircraft on directly opposite tracks and due to the larger turn required by the interceptor to achieve the safety vector heading.

The following example illustrates these aspects of the night interception sequence. The diagram shows the sequence of positions of the two aircraft on the controller's display and the sequence depicting the intercepting fighter turning to follow the intruder for the attack sequence.

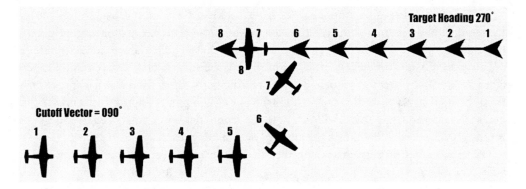

Diagram of sequences of position for an aircraft in a Head On interception.

Lead Collision Night Interception. The sequence proceeds as follows, depicted in the diagram below:

2200—The track of the inbound is initiated and an initial intercept vector of 40° with an altitude of 5,000 feet at a speed of 220 knots is assigned to the intercepting fighters.

2201—The controller directs a right turn to 90°. This is the cut-off vector which will move the fighter from its initial location to the anticipated vicinity of the target aircraft by the most direct path.

2201—The controller informs the interceptors that the target aircraft's track is 180° at a speed of 180 knots.

2202—The controller directs the interceptors to an altitude of 2,500 feet, and informs the interceptors that the target is at 3,000 feet altitude.

2206—The controller directs a speed reduction of the fighters of 20 knots.

2207—The controller directs a right turn to

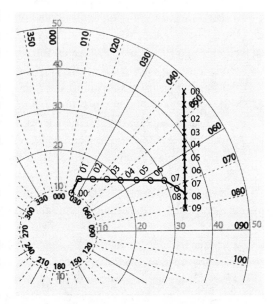

Diagram of an interception using Lead Collision.

140°. This is the safety vector, which will take the interceptor across the track of the target, in this case with a 40° crossing angle.

2207—The controller asks the interceptor lead if the altitude is adequate.

2208—The controller directs a right turn to 180° and informs the interceptors that they are on the target's heading. This is the on-course vector, which places the fighter directly behind the target.

2209—The interceptors report radar contact.

Notice that in the example intercept, all five key pieces of information are communicated to the interceptors.

These tactics and procedures from the late World War II time frame had evolved during that conflict and matured to match the equipment of the time, i.e., piston engine aircraft firing guns and the radar controller having a display showing the radar returns of a single radar, with the history of the aircraft positions having been drawn on the screen by that operator.[12] As with the other topics to be discussed in the book, these tried and true procedures from the mid–1940s were then adapted for the newer equipment and its performance that became available during the Cold War. Both the older and newer equipment were used to accomplish essentially the same basic concept of providing defense.

The preceding illustrations are for the intercepting aircraft at 280 knots and the target aircraft at 200 knots. In the early Cold War period of the 1950s, these piston engine–powered aircraft were replaced with jet aircraft that could travel at around 500 knots to intercept an attacking jet at around 400 knots. This combination resulted in a closing speed of 900 knots, or 15 nautical miles per minute, almost double that of the conventional aircraft example. This led to the need for airborne radar capable of detecting an inbound aircraft at greater range to provide a similar amount of time for the final maneuvering of the interceptor. The first radar developed to assist in this problem was a radar with a beam radiating directly in front of the fighter which would present the pilot with the range to an object to the front. In the early jet fighter aircraft following World War II, the interceptor pilot would be vectored by the GCI controller onto the trail position of a target, then turn on the nose radar to home on the tail of the target. Once the target had been located in this manner, the interceptor would attack at the rear of the target aircraft, beginning to fire the guns within the range of 600 yards down to 200 yards.

In the late 1940s development work began on a new technique to improve the lethality of the interceptor aircraft. The Lead Collision course was emphasized to allow the radar aboard the intercepting aircraft a larger target to track. The larger target for the radar was the side view of the target aircraft, compared to the smaller radar return available from the rear view of the target. This work resulted in more powerful and sensitive airborne radar that could detect the target aircraft at longer range. Also developed for these improvements was the 2.75-inch diameter aerial rocket. The rocket allowed attacking the target aircraft at a greater distance from the interceptor aircraft. As with the previous guns, these rockets were unguided and initially traveled in the direction the fighter was moving at time of firing. Their advantages, compared to the gun projectile, were that they carried a small explosive charge to cause greater damage to the target aircraft, and they could cover the distance between interceptor aircraft and target in a shorter time, decreasing the time available for the target to maneuver out of the way. The new radar was mated with an analog/digital computer to predict the firing point so that the aerial rockets could be fired at the side of the target, preventing the need to maneuver in behind the target.

Fighter aircraft of this period carried as many as 108 unguided rockets. These rockets were fired forward of the fighter and covered an area the size of a football field. To fire these rockets, these aircraft incorporated a collision-course fire control system to place the rockets ahead of the flight path of the target aircraft.

Aircraft	Range	Speed	Weapons	Fire Control System
F-86D	290 nm	603 knots	24 rockets	E-4 (APG-37)
F-94C	700 nm	556 knots	48 rockets	E-5 (APG-40)
F-89D	850 nm	552 knots	104 rockets	E-6 (APG-40)

Table of mid–Cold War fighter interceptor aircraft capabilities.

The D model interceptor version of the F-86 was substantially redesigned from the previous air-to-air models, having only about 25 percent commonality with the previous models. The redesign was primarily to incorporate the larger APG-37 radar, the lead collision course computer, and the belly pack holding the 24 Folding Fin Aerial Rockets (FFAR) which popped downward from the belly of the plane for firing. The interceptor version was powered by the General Electric J47 producing 5425 lbs. of thrust. Over 2500 of this variant were built, making it the most numerous of the F-86 models. At the peak of its service life the F-86D equipped 20 wings of the USAF's Air Defense Command.

Also redesigned from its earlier F-94B predecessor, the Lockheed F-94C was upgraded and became a truly all-weather interceptor with the inclusion of de-icing equipment. The redesign included the more powerful 8,750-lb. Pratt & Whitney J48, incorporation of the Hughes E-5 radar and fire control system (FCS), and a total of 48 of the FFAR rockets: 24 in brackets that popped out around the nose for firing, and an additional 24 in wingtip pods.

The D model of the F-89 Scorpion was primarily redesigned from the previous C model in the weapons system. The guns were replaced by 104 FFAR, which required the Hughes E-6 fire control system performing the collision course guidance for launching the rockets. The Northrop F-89D was considered the nation's first true all-weather interceptor, since it was the first aircraft to have all the desired equipment to operate in bad weather.

AIR-INTERCEPT PROCEDURE (ROCKETS)

The rocket-equipped interceptor aircraft was vectored by GCI onto a collision course directly abeam the inbound aircraft; this was called the 90° beam attack. This collision course was intended to bring the interceptor by the most direct route to a point in the sky where the interceptor's fire control radar could acquire and track the target and provide guidance to the point of launching the aerial rockets. The radars of this era could detect and track an aircraft at approximately fifteen miles. After the interceptor's radar was able to track the target, the Hughes E-4 FCS, or similar, provided a steering circle symbol, which the pilot was to keep centered in the screen by controlling the aircraft. When the computer determined that the interceptor was 30 seconds from the firing point, the presentation on the display changed to a small steering circle in the center of the display and a steering dot which the pilot was to keep centered in this circle. Again at 10 seconds from the firing time, the display changed symbology to a flat line with a steering dot. The pilot was to overlay the steering dot on the line, which indicated that the interceptor was at the same elevation as the target and pointed in the correct direction of the collision course. At this time, the computer inserted a small offset in the aircraft heading to account for the fact that when launched, the rockets would cover the remaining distance to the target faster than the aircraft would. After the firing of the unguided rockets, the system displayed a symbol to the pilot indicating for him to initiate the breakoff maneuver. These rockets took the form of the Mk 4 rockets in the F-86D, F-94C, and F-89D. These rockets had a range of 3700 yards or 1.8 nm. The above procedure for attack with the unguided aerial rock-

Diagram of sequence of aircraft positions in the 90° beam attack used with aerial rockets.

ets was an improvement over the previous attack with guns. This intercept did not require the fighter to enter the trail position of the bogey aircraft. The rocket weapons could be fired from this beam approach, or from a head-on approach, in addition to the traditional approach to the rear of the aircraft to be attacked. The rockets were launched from a distance of 1.75 miles from the target, after which the interceptor aircraft could maneuver away from the target and hence lessen the danger of approaching within range of the defensive weapons of the bombing aircraft.

Deploying Ground-Based Point Defenses

The initial response within the U.S. to the perception of Soviet aggressive intentions included the deployment of antiaircraft artillery batteries around the key population, industrial, and military centers and along the likely approach paths to those centers. These Anti-Aircraft Artillery (AAA) batteries, of the Army's Antiaircraft Artillery Command, of the time consisted of a search radar, a tracking radar, a control room where information was received and orders initiated, and typically four guns, each with an individual crew. The following map shows the locations of the 23 sites where by 1951 the U.S. Army had deployed 64 battalions of AAA.[13] These sites included batteries of 40mm, 90mm, and 120mm guns. The M33 Fire control system was housed in a trailer containing the equipment and operator stations. This radar control room included an early computer to calculate the needed aim point of the guns based on the target movement as sensed by the tracking radar. The track-

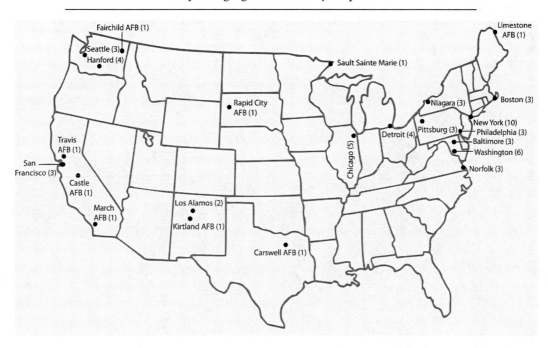

Map of U.S. antiaircraft artillery unit locations in 1951.[14]

ing radar of this M33 system replaced the optically aided human trackers of the World War II equipment. The tracking radar worked in inclement weather by seeing though clouds, whereas the optical sighting required by the previous tracking operations was primarily useful in relatively clear daylight conditions. These sites were manned around the clock by crews not to be more than 20 minutes from their duty stations at any time. When early warning of an unidentified aircraft was received, these crews were recalled to their stations to update the operation centers' tracking map and status boards with the latest information from the collection of radars, and visual observations. If a report was of a particularly threatening nature, interceptor aircraft were dispatched to a particular area to report from that perspective. The early warning element of this reporting system consisted of observation stations on the coasts of Alaska as well as radar sites as far north as Maine. These sites would report via radio to their local headquarters, which in turn would quickly pass the report on to Washington, D.C. From there, the most appropriate areas could be alerted.

Only lead collision interception is available with the gun. These gun batteries had to be located in the suburbs around the perimeter of a large population center. It was this concentration of both population and the industry that it served that was the target of the weapons of mass destruction that the defenses were in place against. Additionally, other key target areas such as the atomic research and development centers and the Strategic Air Command (SAC) bomber bases received this protection. During this period, half of the population of the U.S. resided in 67 critical target areas which comprised only 3 percent of the nation's total area.[15]

The preceding maps depicting the disposition of U.S. air defense forces as of early 1952 are provided to illustrate the defenses available in April of 1952, when events combined to put this system on alert for a possible attack. Three events in quick succession were: 1)

Depiction of M33-style gun battery.

An intelligence report was received indicating a possible air attack on the U.S. 2) High-altitude vapor trails were sighted over Nunivak Island, off the Alaskan coast, traveling in the direction of the West coast of North America. 3) Radar stations in the Northeast U.S. reported five unknown aircraft approaching Presque Isle, Maine. The Alaskan report was sent by radio from the remote outpost to the communications headquarters in Fairbanks, from where it was sent on to Washington, D.C. With the report from Maine a short time later, the commander of the ADC Combat Operations Center in Colorado Springs ordered a full alert of air defense units, as shown in the previous maps, primarily in the Northwest and Northeast regions of the continental U.S. Recall of personnel was made at all air defense bases: fighter aircraft units increased the number of aircraft ready for launch, antiaircraft gun crews were brought to full alert, and other units readied to travel if needed.

Fortunately the tracks reported from Alaska disappeared over the Pacific, and the unknown aircraft approaching Maine were identified as commercial airliners well off course due to much stronger winds than forecast. Nonetheless, the situation exercised the defense preparations in a way a planned exercise could not. The reliance on commercial communication means during an emergency was identified as undesirable. The communication between McCord AFB, south of Seattle, and the air defense center at Elmendorf, Alaska, had failed, cutting off the continental defense controllers from any information following the initial report of aircraft. In the case of the airliners, even though they had reported their course changes to the Canadian flight monitoring stations, this information had not been relayed to the Presque Isle radar site. Another situation needing correction was that it took over 30 minutes for the information to reach the Army Antiaircraft Artillery Command. The lessons learned from this set of events was used to modify plans for additional air defense. Specifically, these events influenced both the magnitude and the pace of deployment of the Permanent radar network and the follow on all-weather fighter aircraft.

Although not ready for deployment in 1952, America's first surface-to-air guided missile had been under development since 1944. The first operational missiles, the Nike Ajax, were initially deployed to begin replacing the guns discussed above in 1954.

The principal objectives of the surface-to-air missile development were to develop a weapon that had greater range and greater speed than the previous artillery weapons used, and for the weapon to be able to react to maneuvers of the target aircraft after the weapon was launched. This latter requirement created the need for guiding the projectile between the time of launch and the time of impact with the target aircraft. The different methods available for this missile guidance are:

Pure Pursuit—By continually pointing the nose of the interceptor at the position of the target aircraft, the interceptor will enter in the trail position of the target aircraft.

Lead Pursuit—By continually pointing the nose of the interceptor ahead of the target aircraft, the interceptor will enter into the trail position of the target aircraft, but by a shorter total path flown than if in Pure Pursuit.

Lead Collision—By pointing the nose of the interceptor ahead of the target aircraft to a point where the two tracks of the aircraft intersect, the interceptor is brought to the immediate vicinity of the target aircraft along the shortest possible track.

These methods of guidance are identical to the methods described for manned aircraft previously.

NIKE AIR DEFENSE

The development of the nation's first surface-to-air guided missile began with a 1944 proposal for how the antiaircraft gun control of late World War II could be extended to control a maneuverable missile to reach a collision point with a maneuvering target. This initiated a nine-year development project to produce the initial entry in the Nike family of air defense missiles. The project was named Nike, after the Greek goddess of victory. The first Nike Ajax battery reached operational status in May 1954 at Fort Meade, Maryland.

The Nike Ajax missile had a total range of 30 miles to an altitude of 70,000 feet. This altitude was higher than the operating ceiling of the threat aircraft at the time: piston engine bombers. The missile reached a speed of Mach 2.3 and carried three conventional explosive warheads, which were detonated by command from the ground system when the missile reached the vicinity of the target aircraft. The missile was radar guided and because of this ground-based guidance, the supersonic missile would be directed to counter any maneuver by the target aircraft.[16] The Ajax missile was intended to take out a single attacking aircraft. Each Nike Ajax missile was 34 feet long, about a foot in diameter, and weighed 2455 lbs.

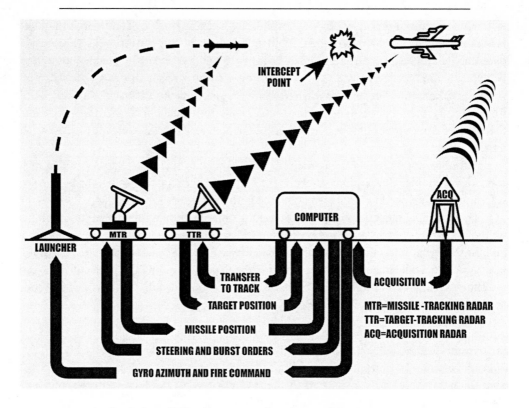

Depiction of Nike Ajax system functions and interconnection.

A total of 6,700 Nike Ajax missiles were deployed at 221 locations. A total of 14,000 Ajax missiles were manufactured.

Common to the two versions of Nike missiles that were deployed were the basing concept and the concept of ground-based control. This ground-based control meant that the sensing equipment, the guidance computer, and the radar system that sent the commands to the missile remained on the ground to be reused on the next missile launch. This single set of controls also meant that only one missile could be launched and controlled at a time from a single battery. A perceived target area was equipped with either four control and launch sites or up to nineteen, depending upon the size of the protected area. A large city such as New York or Los Angeles received the higher number of sites, whereas a more remote Air Force base or atomic research facility received the smaller number of sites. Regardless of the number of sites and hence the number of missiles providing the protection, the sites were placed in a circular pattern around the perimeter of the protected area. Each site consisted of three sub areas: the fire control area, the magazine and launcher area, and the administrative area.

The fire control area could contain the sensing radars and the computer to direct the tracking operations during a launch operation. The magazine and launch area contained the underground storage shelter for the missiles, the fueling support facilities, and the missile electronics maintenance facility. The administrative area contained the unit's command and staff buildings, the barracks for the site crew, dining and recreational facilities. Normally the launch facility was at least 1000 yards from the fire control area, and often the administrative and the fire control areas were adjacent on the same piece of property. The radars also could be located some distance away, such as on a nearby hilltop, if that location provided better coverage for the radars. The multiple batteries located around the area to be protected allowed for different states of readiness among the separate units. This allowed for one unit to be on alert, while an adjacent unit was undergoing training or maintenance.

The second in the Nike family, the Hercules missile, had a total range of 90 miles to an altitude of 150,000 feet. The missile reached a speed of Mach 3.6, was radar guided, and carried a W31 (20 KT) nuclear payload or a conventional high explosive warhead.

There were normally four radars for each Hercules site: acquisition radar, target-tracking radar, target-ranging radar, and missile-tracking radar. A number of configurations were used, primarily depending upon the distance from the battery location to the nearest Air Force radar. The acquisition radar would acquire an inbound target, with cueing information provided by an Air Force radar or a local search radar, the HIPAR. The target-tracking radar (TTR) tracked the target. The target-ranging radar (TRR), which had not been present in the Ajax system, provided a continuous range to the target. The missile-tracking radar tracked the missile in flight. All of these values were fed into the computer that determined the missile steering correction, which the missile-tracking radar would send to the missile as steering commands sent 500 times per second.

The Hercules missile was intended to take out a formation of attacking aircraft. Each Nike Hercules missile was 41 feet long, about 30 inches in diameter, and weighed 10,560 lbs. The first operational site with the Nike Hercules missile was in June 1958 at Montrose Beach

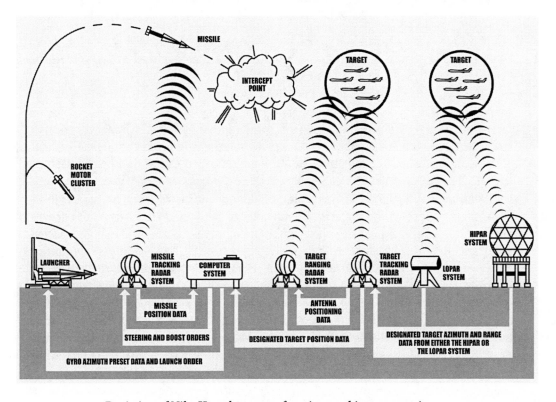

Depiction of Nike Hercules system functions and interconnection.

on Lake Michigan, a short distance from downtown Chicago. A total of approximately 25,000 Hercules missiles were manufactured.

NIKE SYSTEM COMPONENTS

Acquisition Radar: This could be one of High Power Acquisition Radar (HIPAR), Low Power Acquisition Radar (LOPAR), or Alternate Battery Acquisition Radar (ABAR).

High Power Acquisition Radar or AN/MPQ-43
    Frequency: 1350 to 1450 MHz (L Band)
    Range Max: 200 nm
Low Power Acquisition Radar
    Frequency: 3.1 to 3.4 GHz (S Band)
    Range Max: 125 nm
One choice for ABAR was AN/FPS-71.
    Frequency: 1220 to 1350 MHz (L Band)
    Range Max: 200 nm
Target Tracking Radar:
    Frequency: 8.5–9.6 GHz (X Band)
    Range Max: 113 nm (presentation range 200,000 yards)
Target Ranging Radar:
    Frequency: 12 GHz (Ku Band)
    Range Max: 100 nm
Missile Tracking Radar: virtually identical to TTR above.
    Frequency: 10 GHz (X Band)
    Range Max: 114 nm

These radars were connected to the electronics of the site contained in two trailers: the Battery Control Trailer and the Radar Control Trailer.

There were five people located in the Battery Control Trailer (BCT): the plotter, the communicator, the acquisition radar operator, the Battery Control Officer (BCO), and the computer operator. The communicator was responsible for communicating with external radar site(s), the SAGE center, or the Army Air Defense Command Post (AADCP). One of these (and only one as configured) would communicate the location of the target to be tracked by this battery. The plotter would record the commanded target on the Early Warning Plotting Board (EWPB). The acquisition radar operator would slew his cursors to the bearing communicated and to the maximum range of his scope. When a target became visible on this bearing he would slew the range cursor onto this target and keep both his angular cursor line and range circle on the target as best he could. This minimized the time the tracking operators would need to actually track the target. Upon command from the Battery Control Officer, the acquisition radar operator challenged the target with the identification friend foe (IFF) feature; no IFF response indicated an unknown target. When the tracking operators in the Radar Control Trailer (RCT) achieved lock-on to the target, the acquisition radar operator was done with that target and could accept another.

The computer operator controlled two electronic plotting boards: a polar plot, which had one trace for the target and one trace for the missile, and two height traces, likewise one for target and one for missile. The two traces of the polar plot could be assigned to either target or missile as convenient. After lock-on by the MTR and prior to missile launch, these plots recorded the position of the target and the predicted intercept point. These plots indicated the position of the target and the missile based on values provided by the computer, after launch of the missile.

The Radar Control Trailer was also the workplace of five people: the Horizontal tracker, the Range tracker, the Vertical tracker, the Missile Tracking Radar operator, and the Tracking Supervisor. The three tracker operators worked at the tracking console.

After a target was identified by the Acquisition Radar Operator in the Battery Control Trailer, the target appeared in the center of the square screen on the tracking console. The horizontal and range trackers slewed their cursors to find the target. The horizontal and the range cursors had to be close to the target before the vertical operator began to adjust his cursors. Each of these operators viewed a radar signal on their scopes, initially buried in the "grass," or noise caused by reflections from nearby terrain or obstructions such as hills. By working the tuning controls, the operator could reduce this target to a narrow spike in the signal above the grass level. With further tuning, the target became a separate blip in a region separated from the grass background. At this point the target could be locked onto and tracked automatically by the radar.

Following launch of the missile, the steering commands to the missile were sent by the Missile Tracking Radar by altering the pulse modulation of that radar signal.

The members of the operational crews, such as in these trailers, could come from other backgrounds, such as antiaircraft artillery. An operational Fire Control crew was trained by the platoon sergeant. Many people would become cross-trained on the different positions within the platoon. Recruits for the technical positions would attend the school at Ft. Bliss, Texas, prior to assignment to an operational site. A complete crew would train together as well as be on-duty together. The entire crew would undergo retraining with the

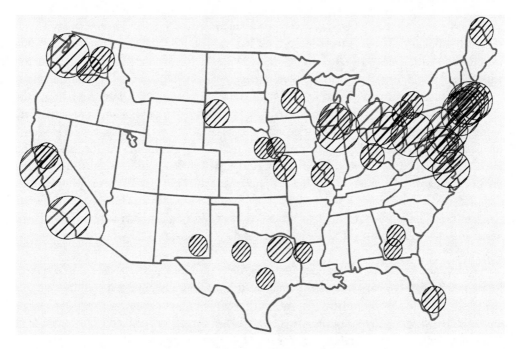

Map of Nike missile sites in the United States.

change of any of the members. On a rotational basis, one crew would be on twenty-minute alert. That is, they could be elsewhere in the immediate vicinity of their station, but on signal they would have to be at their duty station within the twenty minutes. Another battery (site) could be on a longer alert status.

The Nike missile sites were set up in groups called Defense Areas. Each Defense Area protected an urban/industrial area or a national defense site.[17] The national defense sites were Strategic Air Command bases, the Hanford Atomic facility in Washington State, and the naval shipyards around Norfolk, Virginia. The following map shows the locations of the Nike missile defense sites as of 1965. Not shown in this depiction are eight sites in Alaska and four sites in Hawaii. The bulk of these sites protecting the continental U.S. were in operation between 1954 and 1974.

Each of these clusters of missile sites was associated with an Army Air Defense Command Post (AADCP). The AADCP was the post of the Army Air Defense Commander for the particular defense area, say around an industrial city complex. The staff of this command post used a computerized system to systematically handle the multiple tracks operating within its area. This command post assigned hostile tracks to the individual missile sites within the defense area for engagement by a Nike air defense missile. These assignments would be made based on the priority of a specific threat, its location, and availability of firing batteries. Status from the fire units allowed the staff to observe progress of the fire units against assigned tracks, and enabled decisions to bring more units to readiness without delay. A key element of the oversight was to immediately issue hold-fire or cease-engage commands to a particular firing crew if they were preparing to attack a friendly aircraft. Safe passage corridors were established for each area to indicate the agreed-to flight paths of friendly aircraft, to help classify aircraft in the vicinity.

The AADCP received radar returns from its own radars, normally a search radar and at least one height finder. These radar returns were presented as a PPI sweep display at several of the operator console types described below. The AADCP exchanges only digital information with the SAGE Direction Center (DC), described in a following section, and with the missile batteries described previously.

The central computer of the AADCP maintains a common data base from which the symbols for each of six types of interactive display consoles (listed below) are generated. Some of the consoles could modify one or more fields of the common data. The goal was to match the information, position, and motion of each target symbol with the information in the radar returns. The position and motion of the symbol could be controlled by the SAGE data delivered to the AADCP, manual inputs generated by AADCP console operators, or tracking data delivered to the AADCP by a firing site Target Tracking Radar. Because the missile-firing site would have to locate an assigned target and track it with its own radar, the location and the motion of the tracking symbol for that target had to be the best possible.

As with other computer-based systems in the early-mid Cold War period, the system located at the AADCP was intended to present to the several operators using the system a single pictorial display of information from different reference schemes. The different reference schemes were: the search radar reference of information received from the SAGE Direction Center; the local reference of the AADCP's own radars; and the reference of the firing units' individual tracking radars. It was the Surveillance & Entry operators who brought the information together, but it was the Tracking Operators' adjusted location of a symbol that was passed to the firing unit to locate and track with its own radars.

Surveillance and Entry Console (S&E)

This console established, or received from SAGE, the initial coordinates of a track object. The screen symbol overlaid the PPI sweep of AADCP search radar returns. Through a number of buttons on the panels, the operator could change the underlying settings, which would control the symbol seen by all other console presentations. With a joystick panel, the operator could adjust the location of the symbol to align it to the underlying radar return, and establish a direction and rate for the symbol's movement.

The operator of this console managed the state and assignment of all tracks the AADCP system was monitoring. He might assign a track to the RHI operator, then a tracker console operator, to assure that the symbol's position and movement agreed with the local radar presentation. Separate from the main video screen display was a status panel that showed the status of all the possible track channels (48) with indicator lights.

Range Height Indicator Console (RHI)

Here the screen symbols overlaid the radar returns of the height finder radar, with a sweep line pivoted about the lower left corner of the screen and sweeping with the motion of the nodding height finder radar antenna. The antenna swept through a sector of -2° elevation to 35° elevation. With a control wheel containing coarse and fine buttons, the operator could adjust the location of a range-height cursor symbol relative to the underlying radar return. Upon completion, this adjusted height value was entered into the common database that controlled the display of all track symbols in the system. The operator sequentially examined each track assigned and updated the height of the overlying screen symbol as appropriate.

The height value set by the RHI console operator was crucial in the operation in that the search radar determined the target's slant range and azimuth, and the precise height was required for the computer to determine velocity over the ground for proper presentation on the system displays and for correct information to be sent to the firing unit.

Tracking Console (TO)

This console had very similar hardware and function of the S&E console; some buttons were missing, as the tracking operator could not change all data items. The screen symbols overlaid the PPI sweep of AADCP search radar returns. Through a number of buttons on the panels, the operator could change the underlying settings, which would control the symbol seen by all other console presentations. With a joystick panel the operator could adjust the location of the symbol to align it to the underlying radar return, and maintain or correct a direction and rate for the symbol's movement. The operator sequentially examined each track assigned to that console and updated the position of the overlying screen symbol as appropriate.

The role of the tracking operator was to assist the S&E Officer with keeping track symbols assigned to that particular console up to date with the image of the underlying radar return.

The three previously described consoles in the AADCP were used to initiate, modify, and terminate the tracking objects stored in the computer memory of the AADCP's system. The results of these tasks were to have at any one time the best representation of the air situation possible, from the available raw information, to enable the Air Defense Commander to make best use of resources to eliminate attacking enemy aircraft. The assignment of a specific missile firing unit to a specific hostile unit occurred by the actions of the operators of the following tactical consoles.

Tactical Monitor Console

This console had a larger video screen that did not display radar returns; only symbols were displayed. Additional background symbols were displayed showing the location of fire units. This console and its operator were associated with up to eight fire units. A Display Control Panel contained switches to allow the operator to select from all system tracks the track or tracks to be displayed. This console also included switches to connect the operator directly with other stations via telephone so that he could quickly communicate instructions and clarifications.

Track symbols in this display format, called Tactical Display Equipment (TDE), contained depiction of all elements of the track's storage. On the top, bottom, left, and right sides are sets of from one to four dots. On the right side the number of dots depicts the altitude block of the track (in 20,000-foot blocks). On the bottom the number of dots depicts the identity of the track (Friendly, Hostile). On the left side the number of dots indicates whether the track is assigned to a firing unit or not. And on top the number of dots indicates the number of aircraft in the track formation (One, Few, or Many). The center of the symbol is a dot indicating the position of the track. The position dot could have a vector line showing the direction of movement of the track. The position dot was generally in the middle of the track number. The entire symbol was surrounded by a circle after being assigned to a firing unit for attack.

The Tactical Control Panel portion of this console allowed the Tactical Monitor to associate any of the track channels in the system with one of the eight fire units under his control. Within this panel were buttons to send commands to any of the eight fire units and indicator lights to display each fire unit's status. Through a sequence of these button presses and display indications, the Tactical Monitor performed the important step of correlation of a fire unit's receipt of an assigned track as indicated by subsequent fire unit action against the actual intended target represented by the selected track symbol.

Friendly Protector Console (FPC)

This console was very similar to the Tactical Monitor Console in the display capability of system tracks. The Tactical Control Panel was smaller in that it only allowed for sending Hold Fire alerts to the Tactical Monitor. This console also included switches to connect the operator directly with other stations via telephone so that he could quickly communicate instructions and clarifications.

The role of the operator of this console was to monitor the system track symbols and take action when he detected that a firing unit was preparing to fire at a friendly aircraft. This would be indicated when a track representing a friendly aircraft was assigned to a fire unit with an Engage command, or when a fire unit established tracking of an aircraft represented by a track symbol with the friend classification. In either case, the operator initiated a Hold Fire notification to the Tactical Monitor Console. The status of the assignment to the firing unit at the Tactical Monitor Console determined whether the Hold Fire command was immediately sent to the firing unit or whether the Tactical Monitor had to perform a switch press to issue the command. The Tactical Monitor could modify the command to Cease Engage.

Army Air Defense Commander and Air Defense Artillery Operation Officer Consoles

This console was very similar to the Tactical Monitor Console and the Friendly Protector Console. The Tactical Control Panel was different in that it contained a set of Battery Status Indicators, each of which showed the status of an individual battery. Each indicator showed that the battery was ready for action, tracking a target, or out of action. Additionally, there were switches to set the Weapons Control Status, the Air Defense Warning condition, and the Operating Mode of the system. The Operating Mode affected how commands received from the SAGE Direction Center were passed down to individual fire units. The Weapons Control and Air Defense Warning conditions must be communicated to all operators at all sites, as they affected the response by the fire units. This console also included switches to connect the operator directly with other stations via telephone so that he could quickly communicate instructions and clarifications.

The operator of this console supervised the surveillance and tracking operations, and the tactical operations of the AADCP as well as the function of the missile firing units. If there were noticeable differences between the location of tracking symbols and the location of the battery tracking symbols, for instance, the operations officer may consult the S&E Officer to implement corrective action. Or if actions by firing units seemed not to follow the expected sequence, he might issue orders to the Tactical Monitors, the Friendly Protector, or Battery Control Officers for implementation. The Tactical Control Panel was set up to allow observation of the overall activity of the firing units. If, for example, it was observed either that one firing battery received too many tracks to engage or too few, the supervising officer could communicate with the Tactical Monitors to make a change. This is called fire distribution and was a fundamental mission of this entire set of operations.

Fire Unit Console

This console was located in the Battery Control Trailer adjacent to the PPI scope for the fire unit's acquisition radar. Track symbols assigned to this fire unit were visible on the Acquisition Radar Operators scope, overlaying the battery's acquisition radar return. Symbols displayed on the radar scope were: targets tracked by another battery, friendly aircraft, and hostile targets. The console contained a spring-loaded switch to allow the BCO to temporarily display the symbol of the target being tracked by the Target Tracking Radar. This symbol could not be viewed at the same time as the symbols received from the AADCP.

The Command and Status Panel contained switches and indicator lamps which allowed the BCO to send signals to the AADCP and displayed the commands received by the fire unit. Commands sent to the fire unit were Remote (Engage the specified target), Hold Fire, or Cease Engage. Each of these commands was acknowledged by the BCO by hitting the

Acknowledge button. After the tracking crew had achieved a lock-on of a target, the BCO pushed one of the One, Few, or Many buttons to report the size of the raid he was about to attack. At the conclusion of a missile attack, the BCO pressed one of the Effective or Ineffective buttons to communicate the result of the attack. Two other button-indicator combinations allowed the BCO to communicate overall status of the firing unit. He would press the Local button to communicate the site was ready for action, and the Out of Action button to communicate that the site was for any reason unable to attack a target.[18]

SITE READINESS

At least one firing location out of each defense area was on twenty-minute status at any time. This meant that the missiles could be launched within twenty minutes of notification. While on twenty-minute status, checks of the equipment were done every six hours to ensure all was ready. A battery would consist of two Fire Control crews who would alternate twenty-four hours on duty and twenty-four hours off duty during the period that the battery was on twenty-minute status. Similarly, there were two Launch Control crews alternating in the launch control area. There was also a one-hour status and a three-hour status. Rotating among the several batteries of a defense area, one battery was typically on twenty-minute alert status for one week out of a month. This same battery was on one-hour status for another week and in three-hour status for the remaining two weeks of the month. Most training occurred during the week on twenty-minute alert, since all members of the group were present during that time. Each battery of a defense area was in one of the following status levels all the time.

White Status—Not ready. Most equipment is powered to a standby level and the crew is not present. An Early Warning Plotting Board (EWPB) operator may be present to monitor communications and annotate the plotting board with tracks accordingly. If there is no early warning element for this battery, i.e., no AADCP or SAGE Direction Center to provide thirty minutes' warning, the acquisition radar may be powered on and used to keep the EWPB updated with activity in the area.

Blue Status—Preparing to fire, no target assigned. The crew manned the operational positions, all equipment was powered up and checked out. A ready missile was selected for possible firing.

Red Status—The crew was at their battle stations, a target was assigned, a missile selected, ready to fire. The missile tracking radar is locked on to the selected missile. A target is assigned and the Target Tracking Radar is locked on to it. Information is flowing to the computer and the computer is providing its outputs.

At the launching site, some distance from the fire control area, the missiles were stored in an underground magazine where they were protected and could be tested and worked on out of the elements. A missile would be raised on an elevator to the level of the launcher area for firing. The missile could be moved away from the elevator a short distance on the launching rail, so that four missiles could be in launch position. The missile could also be fired from the raised elevator position. In the launcher area was a Launching Control Station, which could be in a trailer, hence the term Launch Control Trailer (LCT). There was a separate launch section for each set of four launchers in the battery. Each section had a Launch Console with an operator. The Launch Control Officer directed the launch control crew. If a missile were launched in response to an attack, launch section crewmen would immediately bring an additional missile to the launch level from the underground magazine, and make it ready for subsequent launch. This would take about two minutes. A firing sequence might be to fire the missile sitting on the elevator first, then, while the elevator was being used, to fetch another round, another missile on the launcher rail was fired, after which the new missile from the magazine would be ready. During a real attack, the battery was expected to fire most or all of its missiles in a short time because the most common attack scenario envisioned was for a formation of enemy bombers to attack.

The sequence of actions a battery site would go through would be the following. A message received at the Battery Control Trailer (BCT) would alert this site for a potential engagement. If this battery was already at twenty-minute status, then the entire crew would be at their stations very quickly; otherwise, they would be assembled, and a siren would direct them to assume their operational stations. The crewmen would each perform checks on their equipment and report to the BCO: Acquisition ready for action, Computer ready for action, TTR ready for action, MTR ready for action. The BCO changes the setting of the readiness and status lights in the BCT, the Radar Control Trailer (RCT), and the LCT, from blue to red. The MTR was locked onto the selected missile's beacon signal, i.e., the MTR is tracking the location of the missile, and this information is being sent to the computer. The BCO reports to the AADCP that the battery is at battle stations. The battery is now at five-minute status.

The Early Warning Plotting Board (EWPB) operator, in the BCT, begins calling out and plotting on his board the position of aircraft tracks reported to him by the AADCP. When these tracks appear on the acquisition radar operator's scope, also in the BCT, they are either surrounded by the friend symbology or the foe symbology as set by the AADCP operators. When engagement is directed by the AADCP, one of the targets with foe symbology also contains the symbol directing this battery to attack that target.

Although the BCT crew can observe the location of the assigned target on the acquisition radar display, if it is beyond the range of the Target Tracking Radar (TTR), they must wait for the target to close to that range. When within range, the BCO commands: "Designate." The acquisition radar operator slews marker lines over top of the target return and moves a switch indicating to the RCT crew that the target at this location is to be tracked; he has "Designated" a target. The Acquisition Radar Operator continues to adjust this cursor position until the RCT crew is tracking the target. In the RC trailer, the Azimuth Operator presses the "Acquire" switch to move the TTR azimuth to the designated angle. He then announces: "Search." At this point the Elevation Operator uses switches to adjust the elevation angle of the TTR antenna until the target blip is visible in the center of the scopes of all three operators. The Range Operator positions the range gate to straddle the target blip and announces, "Target in the gate." All three of the TTR operators move switches to initiate the automatic tracking mode of the TTR radar. The Azimuth Operator then announces "Track" and sets a switch that communicates to the BCT that the designated target is being tracked by the radar. This switch also initiates the sending of the target tracking data to the computer.

With the computer now receiving data from both the TTR and the MTR, its output is sent to the two plotting boards in the BCT. The plotting boards continually indicate the progress of the target's flight path and the predicted intercept point, if the missile were fired. The computer illuminates the Ready to Fire lamp on the console. The BCO is now in complete control to choose the best time to launch the missile, applying his experience and training. When he is ready to actually fire the missile, the BCO announces a countdown of: "About to Fire, ... 5, 4, 3, 2, 1," before he presses the Fire switch and announces: "Fire." The plotting boards stop plotting the predicted intercept point and continually indicate the progress of both the target and the missile in flight. Following launch of the missile, a Missile Away lamp illuminates on the console. Also following launch, the computer sends steering commands to the MTR, which it sends to the missile, embedded within the radar pulses, based on its comparison of target position and track to that of the missile. A "time to intercept" scale is visible at the computer equipment rack; a crewman would observe and announce in the BCT: "40 seconds to intercept," "30 seconds to intercept," etc. The BCO would monitor the progress of the missile flight and intercept to see that safety is proper, i.e., the missile continues toward the designated target.

All three tracking operators continue to monitor the radar returns on their scopes and the quality of the automatic tracking. If jamming is observed, they may take one or more of the tracking channels out of the automatic mode and manually track the target return. The target range is most affected by jamming. In the Hercules system, the Tracking Supervisor can direct the Range Operator to select the Target Ranging Radar (TRR) range data. The TRR is slaved to the TTR in azimuth and elevation but uses different frequencies from that of the TTR and so may not be affected by the jamming affecting the TTR. By changing the source of the range information to the TRR, the Tracking Supervisor can maintain valid range data input to the computer.[19]

The tracking radar operators would be the first to observe the results of the missile interception; the target blip that the radar was tracking disappears. An operator would announce: "Target Destroyed." The computer operator and the BCO would observe the velocity of the target, as indicated by the plotting board pen movement, to fall essentially to zero very quickly. The BCO would report target destruction, or potentially not, to the AADCP.

The Nike missiles were operational until 1974, by which time the threat from Soviet bomber aircraft was substantially replaced by the threat from intercontinental ballistic missiles.

The Pine Tree Line Radars

With the construction and operation of the system of Permanent Radars well on their way to being complete and operational, the military planners of both Canada and the United States, more or less independently, identified the next shortcoming. Canada determined that it too must deploy radars, primarily to defend the Great Lakes and St. Lawrence Valley areas, which contained a substantial portion of Canada's population and industrial centers. Both countries realized that an attack from the Soviet Union must proceed down the flanks of Canada. The U.S. Air Force was faced with the reality that the Permanent System provided warning time of one hour or less for targets in the heavy industrialized areas adjacent to the eastern U.S.–Canadian border. Canada saw that even if the primary targets of such an attack were on U.S. soil, Canadian areas were threatened either as secondary targets, or as targets of opportunity for aircraft that could not reach a U.S. target. The U.S. conclusion was that additional radar sites were needed in Canada to provide more preparation time for an attack on the U.S.

When U.S. defense planners approached their Canadian counterparts about the possibilities of radars on Canadian soil, they found that the practical needs largely overcame the strongly held views of Canadian sovereignty and the need for independent defense capabilities within Canada. Both countries realized that most radar sites constructed in Canada could also provide advance warning to the U.S. The agreement that ensued called for the country that solely benefited from a radar installation to fund that site, and the funding split for a site that benefited both nations. This agreement allowed a path for the U.S. to obtain the desired additional sites to provide needed warning time.

These sites became known as the Pine Tree Radars or the Pine Tree Line. This consisted of 34 sites which performed both warning and GCI activities. These sites were operational in 1954 and were located on both sides of the U.S.–Canada border. Eighteen of the stations were manned by U.S. personnel and 16 stations manned by Canadians.[20]

The most common type of radar installed at the Pine Tree sites was the FPS-20, shown above (drawing by Paul Roth).

Radar	Type	Range	Ceiling	Frequency
FPS-20	General Surveillance	230 nm	65,000 feet	1280–1350 MHz

Table of Pine Tree radar capabilities.

The following map shows the locations of the Pine Tree radars as of 1954.

As with the previous sets of radars constructed, some additional and replacement sites were added in the next several years. Similarly, as the threat became more and more from intercontinental ballistic missiles, these installations began to be shut down in the mid–1960s, although some remained operational until 1990.

Even as these expansions of the sensing and tracking system protecting the U.S. were being agreed to and funded, other discussions were underway to establish even earlier warning for an attack on the industrial and population centers of the northeastern and north-central areas of the United States. With the Pine Tree radars in place, these U.S. areas were afforded 2 to 3 hours' warning time of an impending attack. This warning time was needed by the defense system to alert specific units for action, adjust additional units to a higher state of readiness, and possibly relocate units to a more favorable position to blunt a specific attack. Being debated, both by the technical experts and the military planners of

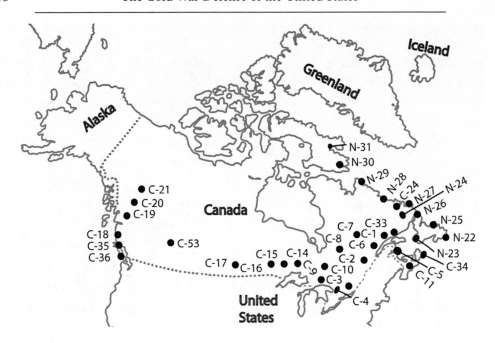

Map of Pine Tree radars in Canada.

Canada and the U.S., was the need for a line of radars in the far north of Canada, approximately 2000 miles from the U.S.–Canadian border. The far north line was known as the Distant Early Warning (DEW) line, because it would provide the U.S. areas with 6 to 7 hours' advance warning.

The assessment of Soviet capabilities and intentions maintained emphasis on the amount of warning time the detection system provided. During this entire period it was known that the Soviets possessed atomic weapons, generally considered to be in the 20 kiloton (KT) range, similar to those dropped by the U.S. on the Japanese cities of Hiroshima and Nagasaki to accelerate the end of World War II. The defenses up to and including the Pine Tree Line considered that Soviet aircraft could deliver such weapons with the Tupolev Tu-4 piston engine bombers to American targets. Air Force leaders never predicted that the air defense system could prevent such an attack, but rather that the defense would reduce the number of bombers reaching their targets and make such an attack very costly to the attackers. This perceived cost to the enemy of 40–50 percent of the attacking force was considered a strong deterrent to the attack in the first place. During the time when the expected bombs were of the 20 KT size, which could devastate a medium-sized city area, such an attack was considered potentially survivable on a national basis. By survivable, this meant that some cities, industrial areas, and military bases were destroyed, but the bulk of the military-industrial might of the nation would survive and could then be employed in the ensuing war.

This concept of survivability ended in 1955 when the Soviet Union achieved its initial success with the thermonuclear bomb. With the 1–10 megaton (MT) weapons then possible, even the number of bombers that might "leak" through the defense system would cause damage that was deemed essentially not survivable. This changed the role of the defense system from protecting the industrial areas so that they might supply the war that was sure to follow, to preserving the ability of the United States to launch a counterstrike. During this time period this massive counterattack capability existed as the Strategic Air Command, or SAC. The deterrent to an enemy attack was that SAC could launch an attack on that enemy, which the enemy would not survive, before the initial enemy attack could destroy sufficient SAC capacity to retaliate.

Some reference distances and flight times to be traveled by a Soviet bomber to reach various U.S. targets:

	Distance	*300 knots*	*500 knots*
Anadyr to Los Angeles	2928 nm	9:45	5:51
Berlin to Boston	3286 nm	10:57	6:34
Murmansk to Chicago	3655 nm	12:11	7:18

Anadyr is located in the eastern Soviet Union across the Bering Sea from Alaska, Berlin is in Central Europe, and Murmansk is in the far western Soviet Union on the Barents Sea and just east of Norway.

In 1954 Soviet long-range aviation, the Dalnyaya Aviatsiya, fielded its initial units of the Tupolev Tu-16 Badger. This twin jet engine aircraft had a range of 3888 nm, a speed of 566 knots, and was equipped for in-flight refueling. The initial deployments of the Mya-sishchyev M-4 Bison bomber took place in 1956. This four-engine jet had a range of between 4320 and 5780 nm, depending on fuel-bombload combination, and a speed of 450 knots. These M-4 aircraft were converted for in-flight refueling in the late 1950s. These perform-ance numbers were a significant improvement over the previous long-range aircraft of the Soviet air force, Tu-4, with a range of 2916 nm at a speed of 301 knots. This combination of speed and range increased the perceived threat of a potential bomber attack by the Soviet Union on the United States, and led Western defense planners to call for the additional warning time afforded by the proposed far-north line of radars.

The DEW Line of Radars

Initiated in early 1954, this line of early warning radar installations was constructed between 1955 and 1957 to be declared operational in July of 1957. There were a total of 6 main sites, 23 auxiliary stations, and 28 intermediate sites. The main equipment on the DEW Line was the AN/FPS-19 search radar, which covered the region from 5000 to 65,000 feet out to 100 miles. The AN/FPS-23 gap-filler scanned above 200 feet, and discriminated targets at less than 125 MPH. The line ran completely across Alaska through Canada, with four stations in Greenland and one in Iceland, near the Arctic Circle, along the 66th-degree parallel. Information was communicated south to the control centers via the "scatter" technique of bouncing radio signals off the ionosphere via a number of rearward communication sites.[21]

The network is made up of the following types of sites:

Main Site:

Housed the same radar equipment as the auxiliary site. Also housed the communica-tion equipment to send warning information south to the control center. There were 6 main sites, 500 miles (800 km) apart. These sites also housed the people who serviced the smaller stations.

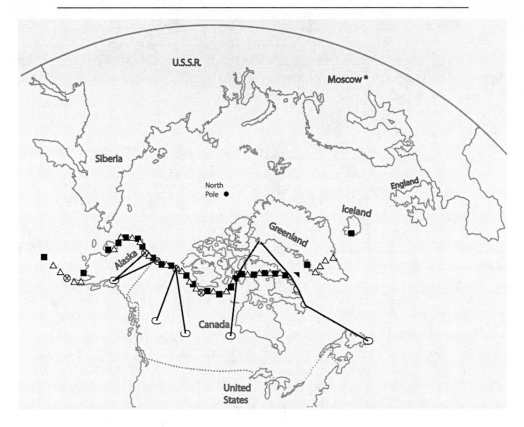

Map of DEW Line radars and rearward communications sites.

Auxiliary Site:

Location of the rotating dish search radar (FPS-19), which had approximately 100 miles (160 km) coverage. One site just barely reached the next site. This site was operated by 10 to 20 men, housed in modular connected buildings. This site normally had a gravel runway for access.

Intermediate ("I") Site:

Located 50 miles (80 km apart), these sites located only Doppler radars (FPS-23), which would sound an alarm at the nearest auxiliary site when something crossed their beam. Intended to be operated unmanned, these sites grew to normally have a crew of 2 to 6.

A detection by one of these radars triggered an automatic photograph of one entire sweep of the radar display, effectively recording the event, in addition to an audio alarm alerting an operator. The operator would then monitor the display to determine the validity of the alarm and send a message via a teletype machine if needed.

Radar	Type	Range	Ceiling	Frequency
FPS-19	Search	86 nm	65,000 feet	1220–1350 MHz
FPS-23	Fence	22 nm		475–525 MHz

Table of DEW Line radar capabilities.

Rearward Communications

To be of value, aircraft detections by the DEW Line radars had to be communicated quickly to the command apparatus in the lower forty-eight states. This communication was unusual owing to the remoteness of the radar sites and in many cases by the rugged terrain along the way in the southward direction. The terrain of Alaska brought about developing solutions to these challenges prior to development of the DEW Line, and applying those solutions to the DEW Line as needed. These developments resulted in what was known as the White Alice Communication System (WACS).

The White Alice Communication System consisted of over 80 individual stations. There were two types of transmission: tropospheric scatter was used for long over-the-horizon links, and microwave was used for shorter point-to-point links. Each site contained two sets of antennas, one to receive the inbound signal and one to transmit the signal outbound to the next site. Each of the sets of antennas could be of either type, depending on the type of the link, tropo or microwave. The tropo antennas were large parabolic surfaces, while the microwave antennas were smaller round dishes.

These combination tropo/microwave links were used for lateral communications along the DEW Line and from the line to the Alaskan White Alice system. For rearward communications, i.e., from the north back to the forty-eight states, ionospheric scatter was used. The tropospheric and ionospheric scatter techniques both directed the radio beam above the horizon so that particles in one of the layers of earth's atmosphere would reflect the energy back into the atmosphere. For each of these atmospheric layers, the reflection was predictable so that a receiving site could be located. A small number of communication relay sites were established in Canada's interior, and in Alaska, to receive these long-distance communications from the main DEW sites. These sites then retransmitted the information further south. When the Ballistic Missile Early Warning System (BMEWS) sites were added to the array of sensing, ionospheric links were added from Thule, Greenland, to Cape Dyer, Nunavut, and from Thule to Hall Beach, Nunavut, to relay information southward. From the southernmost relay sites, the messages could continue south via more conventional methods.

The Rearward Communication ionospheric scatter sites were equipped with IS-101 (FRC-101) radios, which communicated teletype data. This system was not capable of voice quality. The individual radio links implemented with the ionospheric scatter equipment were:

- From Point Barrow, Alaska, and Barter Island, Alaska, main sites to Anchorage, Alaska.
- From Cape Parry, Northwest Territories, main site to Ft. Nelson, British Columbia.
- From Cape Parry, Northwest Territories, main site to Stoney Mountain, Alberta.
- From Hall Beach, Nunavut, main site to Bird, Manitoba.
- From Cape Dyer, Nunavut, main site to Brevoort Island, Nunavut.
- From Brevoort Island, Nunavut, relay site to Resolution Island, Nunavut.
- From Resolution Island, Nunavut, relay site to Goose Bay, Newfoundland.

These links are depicted on the previous map.

The communication links using the ionospheric scatter technique were the longest in the world up to that time. The link from Thule to Cape Dyer was over 600 miles, quite remarkable at the time. The high maintenance needed by the system components and the fact that this system could not provide reliable voice communication led NORAD to search for a replacement for the ionospheric scatter system. These ionospheric scatter sites were closed in 1963 and replaced by satellite communications.

SEAWARD EXTENSIONS

The U.S. Navy had well learned the concept of the radar picket ship in World War II, particularly in the Pacific theater. There a destroyer would be stationed at the extreme perimeter of a fleet of naval ships to utilize its air search radar to detect an air intruder some distance from the main body of ships. This concept evolved to the practice of placing air controllers on this ship to direct fighters in an attack on the intruders.

Similarly, after the war, the Navy had evolved essentially a flying radar station to monitor a sector of airspace and have at hand the necessary people and equipment to direct friendly aircraft to deal with hostile intruders. This evolution had started with a version of the Grumman TBM Avenger, followed by a Navy version of the Boeing B-17, to be replaced by a version of the Lockheed Constellation.

In 1954, these ideas were revived to augment the lines of land-based radar stations over the oceans adjacent to both coasts of the North American continent. Using both ships and aircraft, lines of radar coverage were established out to sea about three hundred miles from the coastline. These seaward stations reported any detections to the network of control centers located on shore. The Atlantic Contiguous Barrier was made up of five patrol areas stretching from Cape Cod to North Carolina. Similarly, the Pacific Contiguous Barrier consisted of five more patrol areas located from Seattle to central California. These stations provided extension to the Pine Tree Line of land-based radars running roughly along the border between Canada and the U.S. The ships employed were World War II destroyers, DE's converted into DER's with updated radar. These DER's began to be replaced in 1960 by the AGR ships. These ships were converted from remaining World War II transport ships to the radar station operation. The aircraft were the WV-2s, based on the Lockheed Super Constellation passenger airliner. The Air Force also used essentially this same airplane and equipment, labeled the EC-121.[22]

With the building of the DEW line farther north, providing additional warning time of a bomber attack, a seaward extension was also called for. The resulting Atlantic Barrier and Pacific Barrier were the result, which began operations in 1957. The Atlantic line from Canadian Newfoundland to the Azores was made up of four stations at a spacing of two hundred fifty miles. The Pacific Barrier ran from the Aleutians to Midway Island, at a spacing of two hundred miles. Each of these patrol areas contained a ship or an airplane, sometimes both.

In 1961, the Atlantic Barrier was either replaced or moved north to become the Greenland-Iceland-United Kingdom, or GIUK Barrier. With the completion of DEW line radars across Greenland, the previous position of the Atlantic Barrier became redundant. This GIUK barrier consisted of two surface ship radar stations, one west of Iceland, and one east of Iceland. A matching set of airborne patrol stations were patrolled by the WV-2 aircraft based at Keflavik, Iceland. Also in 1961, an additional barrier segment of the Atlantic Contiguous Barrier was established about 100 miles east of Key West, Florida, and

80 miles south of Miami. This Southern Tip station provided warning of aircraft approaching the U.S. from the south, primarily from Cuba.

A ship patrolling a barrier station was most often on station for from one to four months while the aircraft would conduct a twelve-hour mission. Every four hours the next aircraft would depart its base to join the revolving set of planes watching these skies, and on the same schedule one would return to base. For both ships and planes, the weather in these patrol areas was terrible. The limited balancing element for the discomfort endured by the crews was that they became quite adept at operating in the nasty conditions.

These radar sensing barriers over ocean waters adjacent to North America were discontinued in 1965. The steady improvements in the land-based radar system, combined with the shifting of the perceived threat from manned bomber to missile, removed the need for this radar coverage.

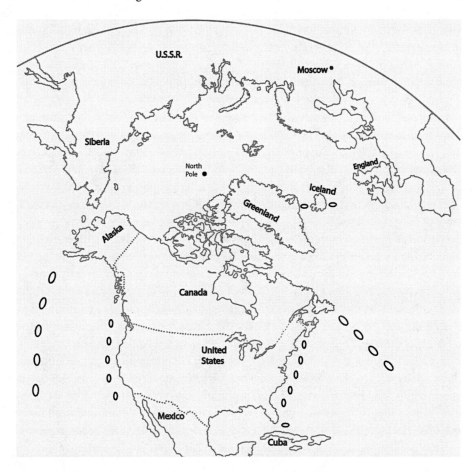

Map showing the location of the Atlantic, Pacific and GIUK radar barriers.

Centralizing the Control of Radars, Fighters and Missiles

As with the early warning radars, the fighter interceptor aircraft, and the ground-based antiaircraft weapons, the communications and control of all of these elements to provide a complete defense system evolved during the time following the heightening of tensions with the Soviet Union around the time of the Korean War.

Through the period of the Permanent and then the Pine Tree radars as the early warning detection of the air defense system, the position of both the incoming hostile as well as the defending interceptor aircraft was determined by these radars. The individual scope operator at the radar site made a determination of the position and direction of travel of the aircraft involved and reported this information to others. The information was then passed by telephone or teletype to other, including higher echelon, stations. Both at the radar controller site and the higher command station, the relayed information was captured on plotting boards so that it could be viewed by others in a pictorial fashion with appropriate landmarks such as coastlines, cities, and fighter interceptor bases permanently drawn for reference. As an attack progressed and the locations of the aircraft involved changed, the information on the plotting boards was erased and redrawn with the newest known information. This arrangement was little changed from World War II and was generally adequate when the aircraft were traveling between 200 to 350 miles an hour. It was also adequate because the volume of airspace monitored by a single radar site could contain both the hostile target aircraft and the interceptors dispatched to deal with the threat.

From 1950 through 1953, several groups of select individuals, mostly made up of scientists and senior military officers, were assembled to study and make recommendations for improved effectiveness of the national air defense system. These groups added weight in favor of constructing the Distant Early Warning line of radar detection. Additionally these groups recommended a national effort to systematize the command and control aspect of the defense system. This concept centered around applying the then revolutionary idea of a digital computer performing many of the detail operations of receiving, managing, and responding to information generated by the lines of radar sensors.

In 1950, the Project Charles study group was convened to study the air defense needs of the U.S. and make recommendations to the USAF. Central to the findings was the need to increase the speed, reliability, and volume of the transmission of information needed to respond to an air attack. By analyzing military exercises, the study group identified that the people handling the different types of information during an attack became overloaded, and additional information beyond that level was significantly degraded. For example, the two people making the initial collection and recording of an aircraft track, the radar scope observer and the plotter paired with him, could do no more than about 10 tracks before saturation. To handle the maximum size of attack at the envisioned future aircraft speeds, the operations room and staff would have to handle hundreds of inbound attackers and scores of defending aircraft. This study largely reinforced the correctness of some developments underway and recommended some adjustments. Major projects underway in the area of air defense were:

1. apply a digital computer to process radar signals and to initiate defensive response,
2. develop a faster, higher-flying interceptor jet aircraft incorporating guided missiles,
3. develop unmanned guided surface-to-air missiles.

The electronic system evolving for the Air Force at the Massachusetts Institute of Technology (MIT) Lincoln Laboratory brought the detection information from the radars into the central location of the digital computer by means of digital communication. The computer was able to do the filtering process of determining which radar detections were the next successive positions of an aircraft already being tracked. To do this, the computer distinguished some reports as redundant if coming from radars with overlapping coverage of another. Other reports were identified as a new target entering the tracking process. Information on known aircraft movements, both civil and military, as well as the status of response forces, also arrived in a similar manner. The computer was then able to compare the radar-detected aircraft with expected movements to determine those potentially hostile. The computer periodically rechecked these track comparisons as the detected aircraft moved through the area. With the replacement of the manual telephone calls previously used to communicate a target to a defense weapons site with the digital communication, substantial time was saved in getting a weapon on its way to the target. The human operator, in the weapons direction section, made the decision whether to assign a manned fighter interceptor or an unmanned interceptor missile to a target. The first task of the pilot of the interceptor was to determine the intentions of the unknown, then to launch a weapon following visual identification, once in the immediate vicinity of the target, if warranted.

The project underway to provide for a much more capable interceptor aircraft, project MX-1554, often known as the 1954 Interceptor, included the ability for the automated system to transmit to the aircraft the location of the incoming target. By generating track information on both the defense aircraft and the inbound, derived from the radar data, the computer was able to constantly update the information sent to the intercepting aircraft. The final phase of the air interception, the launching of the selected weapon, could be tailored by the computer to match the selected weapon, whether gun, rocket, or missile. Another operation of the ground and airborne systems was the return to its base of the interceptor aircraft. This project produced the Convair F-102 aircraft, which later evolved into the F-106.

At the time of this study report, the Nike as well as the BOMARC unmanned and guided surface-to-air missiles were under development by the Army and Air Force, respectively. The development of the air-to-air guided missile was also underway at this time under Air Force project MX-798. This project would produce a guided missile to be launched from a manned interceptor aircraft. This project developed the Falcon family of missiles built by the Hughes Aircraft Company.

This Project Charles study, and others that were to follow, reinforced the concept that the progression in weapons from guns to rockets to guided missiles was correct. For each step in the evolution, the pilot of the interceptor aircraft was able to launch the weapon from greater distance, while having greater probability of success, and needing to maintain a precise positional relationship on the target enemy bomber for a shorter time.

We now understand that the constant redrawing of the air defense picture, as described above, with the newest information, determining additional information based on changes in the reported information, and communicating this information to other stations, is precisely the domain of the digital computer. But in 1950, this was not only beyond the technology of the time, but beyond the imagination of all but a few.

The air defense automation project benefited from an earlier MIT project funded by the Navy. This project developed an early digital computer called Whirlwind. Although this work resulted in computer performance that made it unclear if there was a military application for the system, the lessons learned by the MIT group were invaluable. Furthermore, this Whirlwind (I) allowed testing and evaluation of the broader concepts involved in the air defense problem, while engineers went to work on the performance limitations. The result was Whirlwind II.

The system that evolved eventually became known as SAGE, an acronym for Semi-Automatic Ground Environment.[23] Central to the SAGE system was the MIT computer initially known by the name Whirlwind II. The Whirlwind II, when manufactured by IBM, became the AN/FSQ-7. What became known as the SAGE computer was comprised of some 60,000 vacuum tubes. Early trials to prove the feasibility of incorporating the digital computer to improve the response to an airborne attack were successful. Following this, key elements of the development were to determine how to construct what would become the air defense computer to be efficient, reliable and effective.

A key technological decision made prior to Whirlwind II was to utilize magnetic core memory as the basis of the internal storage of the central computer. The Whirlwind I had used electrostatic storage tubes, one of the performance and reliability issues of the earlier design. Transistors, the foundation building block of all modern integrated circuit electronics, were in existence—barely—but were not adequately available and did not have any history on which to base the all-important reliability prediction.

Input-output buffering was the mechanism to account for the different speeds of the equipment. The central computer could either read or write magnetic drums very quickly, faster than the external circuits could provide or receive the data. The serial communications of the day were 1300 pulses per second, so although it took the central computer less than a millisecond to read an incoming message from a radar site, it took about a second for the communication to get a complete message onto the magnetic drum. In this way the magnetic drum was the temporary storage for the message until the computer was able to read it.

Similarly, the computer would store information on the display drum to draw the display screen for each operator station in the Direction Center. Then the dedicated hardware would read the display drum every 2 1/2 seconds to update each display scope. To increase the efficiency of this process, the computer would only change those words of the output message necessary as the computed information changed with the movement of aircraft, or other changes in the information to be displayed.

These magnetic drums were another key to the success of Whirlwind II. They stored, at least temporarily, most of the information available to the computer. A typical sequence of operations was for the computer to copy a segment of the program into the high-speed "core" memory, then copy other areas of the twelve magnetic drums needed by the program, then do the calculations of the program, then write any of the tables loaded into core memory that had been modified by the program back out to its place on the magnetic drums.

The initial test system consisted of a single search radar and a couple of airplanes to provide targets for the system to track. The testing of this system, its components, and the people who would control and respond to it, evolved in the Cape Cod area of Massachusetts to include more radars, more aircraft, and the other elements depicted in the following diagram to produce the Experimental SAGE Sector.

Twenty-two SAGE Direction Centers (DC) were built in the U.S. following the completion of the test projects around Cape Cod and the initial SAGE installation at McGuire AFB, New Jersey. These centers were in place by 1961. Each of these centers was the central facility of an air defense sector. A SAGE Direction Center could be built either on an established military

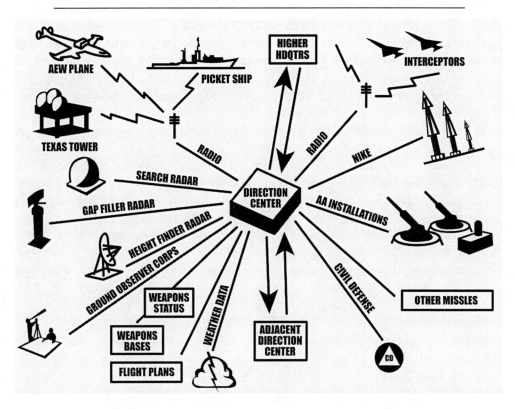

Depiction of SAGE Direction Center and interconnection with the sensor and weapons sites.

base, such as the example of McGuire AFB, or it could be on a small base where the DC was substantially the sole function of that installation. The small base would employ about 850 military personnel, officers and enlisted, and about 150 civilians to support the technical aspects of the operation. If installed on an existing base, the number of people assigned would be smaller, since this facility would be only one of several assigned to that base.

DIRECTION CENTER (A SAGE BUILDING)

The Direction Center is made up of several functions performing the air defense operations: air surveillance, identification, tracking, and weapons sections.

Air Surveillance consists of the Radar Input section, with a countermeasures officer and technicians who communicate directly with the Air Defense technicians at the radar sites, and the Manual Input section, which provides information on flights from air traffic control and other DEW line sites for which there is no automatic input mechanism.

Identification must resolve an unknown track within two minutes from detection. For an unknown to be identified it must be less than 20 nm and less than 5 minutes from the predicted path of an established flight plan.

Tracking monitors all radar contacts or "tracks" in the sector. An unknown track is assigned to the identification team for determination of possible hostile status.

The weapons section is made up of a director, a technician who communicates directly with the operations facilities at alert air bases, and five intercept controllers, each with a technician who communicates directly with the alert air crew following launch. These intercept controllers work out the intercept geometry of efficiently bringing the alert aircraft to the point where either a visual identification of the target or tracking by the aircraft's radar can be accomplished by the crew.

The Senior Director oversees all this activity and makes necessary assignments and decisions; for example, directing that the weapons section initiate investigation of an unknown track by scrambling a pair of interceptor aircraft from an appropriate base. To facilitate these decisions, information is constantly displayed and updated on the status and availability of the radar sites, the communication links, and the fighter aircraft which are all used in the ongoing process.

All of these functions rely on information from the AN/FSQ-7 computer. Most often the operators in the Direction Center building viewed information on a Display Scope station with a number of switches allowing them to place additional information on the display or remove some information from the display.[24]

The Display Scope was a 19-inch diameter cathode ray tube (CRT) called a Charactron. This display device was surrounded by panels of switches for the operator to signal the computer to modify the information displayed on that particular station. Some stations were also equipped with a 5-inch CRT called a Typotron display device, which could display 8 lines of 14 alpha-numeric characters each, with no graphics. This device was used to display status information to the operator, including the reason an operator request had not been completed. Some stations were equipped with a typewriter-style keyboard if that station was to directly input information for storage to be used by the computer. All stations were equipped with a light gun. This was a hand-held device which the operator pointed at a particular symbol on the Display Scope to make the focus of a following action. Two common actions were to cause the detailed status of the object represented by the symbol to be displayed on the Typotron display, or to change status of the object by input from that console. For example, to assign an interceptor fighter to investigate a certain target track, an operator would select an intercept symbol with the light gun.

The present-day computer mouse or touch screen pad is considered a direct descendant of the light gun. The graphical display screen and the dumb terminal display would be direct descendants of the Charactron and Typotron displays respectively. These elements were all available for the SAGE system testing beginning in the mid–1950s. If you have believed that the computer mouse was invented by someone in Silicon Valley in the 1980s for the revolution of personal computers, you would only be partly correct.

The fourth floor of the SAGE blockhouse contained about fifty such operator stations, with several types of console hardware with a switch configuration unique to a specific set of tasks.

These tasking differences were separated into different rooms on the floor. Rooms for surveillance, identification, status input, weapons, simulation, and command performed the operational functions of air defense. A separate room controlled and monitored status of the two independent computers in the building; both received inputs, but only one performed the calculations and provided outputs. By receiving the same inputs as the active computer, the standby computer could be switched to active and begin generating the outputs very quickly, and would be barely noticeable to the operators. This allowed one machine to be undergoing maintenance while the other did the primary work. Before the standby computer could go active, its memory would have to be updated with the status tables from the active computer. This would occur several times a minute if the standby computer was operating.

The simulation group and room in the Direction Center created and initiated training exercises by inserting simulated events into the system. For example, this group could insert into the computer memory some information as if such a message had arrived from one or more radar sites. The various computer programs and other people at the various operating consoles would not know this was a training exercise, until told by the simulation group, and would perform their functions on the information as they were supposed to. Similarly, the simulation group could temporarily alter the status information of an air base, thus forcing the weapons team to select a certain base for the response to the reported unknown aircraft. By altering one type of the system information or another, the simulation group could devise an exercise to test the response, and hence the training and readiness, of practically any individual element of the entire system. Similarly, to conduct a complete system test, friendly aircraft would be included in an exercise of the entire system. A friendly aircraft could fly, say, from an offshore area towards North America. Without knowing this was an exercise, one or more radar sites would detect the plane and make the reports. In this scenario, fighter aircraft would be dispatched and should locate the aircraft. Obviously, the weapons team and the fighter aircraft crew would need to be told of the exercise before the critical time of weapons launch, but this was within the ability of the simulation section. If would be a bad day indeed if such an exercise led to the launch of an unmanned interceptor missile which could not be recalled; this never happened.

The computing hardware was made up of a number of subsystems surrounding the central computer. Only this central computer is capable of calculations based on the instructions fed to it, called the software. These other subsystems primarily implement the low-level details of input or output operations. For example, an output operation would be organized to always send the contents of a specific area of a magnetic drum to a particular Display Scope within the DC. Or the input message from a radar site would always be written to a specific area of a magnetic drum. The computer would read the magnetic drum to input the radar message, perform the needed calculations to update the position of, say, the unknown target aircraft track, and write the new position of that aircraft's track symbol to the proper magnetic drum location for display on the appropriate Display Scope. This use of magnetic drums to store the input or output information until it is completely sent or received is called buffering.

SENSOR SYSTEM AND FILTER CENTER

Each radar image was presented to the common digitizer (AN/FST-2) at that site. This common digitizer, also known as the coordinate data transmitter, translated the radar screen image into digital messages which were transmitted to the SAGE center via telephone lines. Areas of the radar screen were masked from the Common Digitizer by marking out those areas with an amber-colored grease pencil. These areas of the display showed the returns from nearby hills or water returns from beyond the coastline. The information extracted by the digitizer was the center azimuth of a return plus the "run length," or width of the return, and the range of the return. This information was sent as a digital message via a phone line (modem) connection to the Direction Center, the building which housed the AN/FSQ-7 SAGE computer and collection of operators described below.[25]

This input radar information was first handled by the Radar Input Counter Measures Officer and his group of Radar Input Counter Measures Technicians. This team worked directly with the Air Defense Technicians in the Data Management Control Center of each radar site. Together, they collaborated to produce the best information possible on the target. Some areas, particularly when covered by the Pine Tree radars, were covered by more than one search radar; in this case each radar site may display a target for a given aircraft. The SAGE system would compare the multiple reports from separate radar sites and determine that one target track was required. This target was viewed on the Identification Officer's screen where the Surveillance Technician would initiate the system's tracking of this target by determining the course and estimating its speed. Initially this new track was classified as "pending," which initiated a two-minute period in which the Identification Officer had to either identify it as belonging to a known commercial, civilian, or other military flight, or declare it as an unknown. The Identification Officer announced to the facility that this new track was "pending." To identify the new track, the technicians compared the radar track information with flight plans from the air traffic agencies. These flight plans from the agencies were communicated to the direction center by teletype. For a track that was declared an "unknown," because it did not match any known flight plan, the Identification Officer announced to the facility that this track was "unknown." The Senior Director then assigned this track to a weapons team that had interceptor aircraft available for the intercept. The selected team's Weapons Director verified the previous status of interceptor aircraft availability and alerted a fighter unit of the likely use of their aircraft, or ordered available fighters scrambled. The Weapons Director then assigned this intercept to one of the Intercept Directors on that team.

The Intercept Director would manipulate the Display Scope presentation to facilitate display of the target on the screen with other symbols nearby for reference. The target symbol included a vector line indicating the target track direction and speed, and a set of letter characters called the data block, each character indicating an item of status or resources assigned to this track. Other symbols depicted the location of interceptor bases, defense missile sites, and a predicted intercept time for aircraft from bases that had available interceptors on alert. The Intercept Director would pick one of these predicted intercepts, perhaps the farthest from the expected target area, and assign those alert aircraft to this track. The Intercept Director Technician working with this director passed the order to the selected squadron to scramble their two alert aircraft, if this had not already been done by the Weapons Director. After these aircraft became airborne, they were directed by air traffic control to contact their weapons controller on a frequency that put them in radio contact with the Intercept Director assigned to this mission.

The following diagram depicts the display screen that the Intercept Director would see if the inbound track were approaching the northeastern U.S., in this case near Cape Cod, Massachusetts. Depicted is the coastline, three interceptor bases, and the two closest Nike missile batteries, one from the Boston Defense Area and one from the Providence Defense Area. Notice that interceptors from bases at Otis AFB and Hanscom Field would arrive at nearly the same time, truly providing the Intercept Director with a choice.

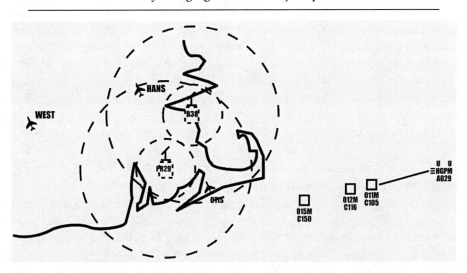

Depiction of SAGE screen presentation showing radar targets and defense sites near Cape Cod, Mass.

Following the selection, the screen symbology would reflect the assignment for both the target and the fighters. The target symbol would show a weapons team had been assigned, a specific Intercept Director within that team assigned, and that interceptor aircraft were assigned. The remaining predicted intercept point would now reflect the Intercept Flight Code Number which the system had assigned along with the time and distance of the interceptors from this point. A new symbol for the fighter aircraft would appear at this stage to indicate the track number and the Intercept Flight Code Number to which they were assigned. This new symbol included the vector indicating the course and speed, and the data block indicating the quality of the tracking, the weapons team that was controlling, and a letter indicating the altitude.[26]

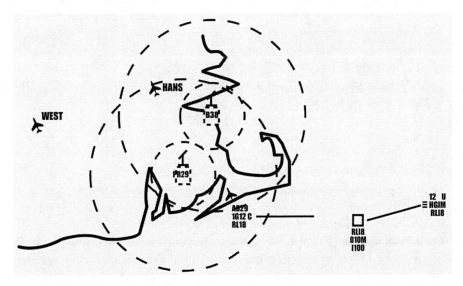

Depiction of SAGE screen presentation showing interceptor assignment on target and defense sites near Cape Cod, Mass.

The Intercept Director's job was to guide the flight path of the interceptor aircraft into a position where the aircraft could locate the unknown target on its own radar, and hence take over the remainder of the intercept by directions provided by the Airborne Intercept Officer in the back seat of a two-man aircraft. To do this, the Intercept Director had the option of directing the aircraft to "Follow Dolly," that is, direct the flight crew to follow the directions transmitted to the aircraft by the SAGE system. Such commands could be carried out by the on-board autopilot and essentially remove the flight crew from the target tracking problem. The system on board the aircraft was one of Hughes Electronics' Fire Control Systems. For example, the MG-13 system of the McDonnell F-101B was comprised of some 200 black boxes and weighed some 2520 lbs.

After the aircraft was en route to an intercept with the unknown aircraft, the Radar Input Countermeasures team continued working to refine the radar information about the target. One or more of the radar sites would be directed to use its height-finder radar to obtain the altitude of the target aircraft. To do this, an Air Defense Technician at the radar site would command the height-finder radar to swing to the azimuth of the target as determined by the surveillance radar which determined azimuth and distance of the target. With the height finder pointing at the proper azimuth, the technician would adjust the nodding antenna of the height finder so that the target was in the center of the antenna's beam. When satisfied with the presentation, he would send the information to the SAGE system by the press of a button. This step of altitude determination resulted in the "semi" prefix in the Semi-Automatic name of the system.

If radar signal jamming was detected or expected, which it was, the Radar Input Countermeasures team would also work on this problem. The location of the target or jamming aircraft could be determined by multiple radars' observations of the received jamming signal, if the target conducted the jamming for any length of time. If jamming was present, it was possible that a specific radar was less affected and could provide the primary information. The transmitter and receiver frequencies of the various ground radars were adjusted as those systems allowed to reduce or avoid the impact of the jamming being done by the target aircraft. To accomplish this, the Radar Input Countermeasures team would initiate a common communication between themselves and the participating radar sites.

Key differences between this operation and that described earlier for radar intercepts using piston-engine fighters are:

1. The operators at the SAGE Direction Center never saw actual radar images; presented on their Display Scopes was a composite of computer-processed information derived from several sources. The current lingo would call this "data fusion."
    a. The target and the defending fighter can be shown on the operator's screen regardless of the distance between them, and without regard to the radar site that produces the best representation of their positions.
    b. The computer can calculate a new approximation of the interception with the previous information on direction and speed for the aircraft, in between updates from the radar information.
2. The computer redraws the situation display with status information overlaid, based on computations using the newest information as it arrives.

3. With the information on this unknown track and the status of the response stored in the memory of the computer, other screens in the Direction Center can display the information as well. Supervisory personnel and the commander can view all such information throughout the Defense Sector to see the "bigger picture."

    a. This assessment is used to make changes to the readiness of response forces. For example, the commander could order additional aircraft brought to full alert status at certain bases.

    b. This assessment is used to communicate up to the NORAD Region, where an even bigger picture is compiled from the multiple Direction Centers reporting to a region. This could lead to moving aircraft to a more forward base, based on the trends observed.

4. All the people in the DC are more focused on their specific responsibility because they can tailor their screen display to remove that which is not immediately pertinent, and then allow the computer to redraw their selected view as the information from the various sources is updated.

DESCRIPTION OF AN ONBOARD FIRE CONTROL SYSTEM

This description is of the Hughes MG-10 system of the Convair F-102 aircraft, although many of the concepts apply to the several systems used on the various aircraft previously described. The MG-10 is not the earliest nor the latest fire control system used during the Cold War period being described. The purpose of the system is to allow the pilot of the aircraft to locate a target from a safe distance away, attack the target by launching weapons, and return the pilot and aircraft safely to its base. The MG-10 system uses a radar, a computer, a flight control system (often called the autopilot), and several supporting components to do this such that the pilot initiates the key actions, but the system is capable of doing much of the second-by-second operations automatically.

This system, an early example of an integrated avionics system, is comprised of six major elements: a radar, a data link receiver, a computer, an automatic flight control system, missile interfaces, and an optical sight. These system components transfer information from one to another to affect the overall system operation. The data link component provides the target range, target bearing, heading for the interceptor to fly, time to go to the destination, and the target altitude. This target altitude is presented to the pilot on a separate panel meter for that purpose. These other values, target range and bearing, heading to fly, and time to go, are used to position the Target Marker Circle, the Steering Dot, and the Time To Go Circle on the radar display to direct the pilot to locate the desired target. Once the radar has been locked onto a target, the radar sends the range, angular position, and angular rate of the target. The computer uses this information to perform the fire control computation, which results in steering commands sent to the automatic flight control system, and information sent to the radar scope to update position of the Steering Dot. The computer also sends information to the missile interfaces to prepare the selected weapons for firing.

Following takeoff of the aircraft from its base, the system receives radio messages transmitted by the ground control intercept (GCI) station directing the interceptor to a position from where an attack may be initiated. This position is called the moving Offset

Point because it moves based on the position of the target. While the aircraft is traveling towards the Offset Point, the flight control system, if enabled by the pilot, controls the aircraft based on the messages being received from the ground. Also during approach to the Offset Point, the radar is searching for the target, also based on the messages received. Upon reaching the Offset Point, the system steers the aircraft to fly a collision course with the target's position. Upon locating the target return in the radar, the pilot initiates the radar lock-on, which initiates the radar's automatic tracking of the selected target. Once a target is locked-on, the system controls the aircraft to fly the correct attack course for the type of weapon selected. This could be either a lead collision course or a pursuit course depending on the weapon selected. The pilot engages the autopilot's control of the aircraft on the attack course by releasing the control grip. During this time when the target is locked-on by the radar and the system is flying the aircraft on the attack course, the system also sends information to the missile interface in preparation for launch. This operation initiated power to the missiles to effect a warm-up, send commands for the radar missile seeker to synchronize with the fighter's radar, and point the missile's radar antenna in the direction of the target. At the correct time determined by the computer calculations, the system launches the weapons. Following launch of the weapons, the pilot is signaled by a symbol on the radar scope to terminate the attack run by taking manual control of the aircraft to change course and avoid effects of the weapons.

The pilot controls the system largely by interpreting symbology provided on the radar scope. While the aircraft is traveling towards the Offset Point under control of the messages received from the GCI station, there are four symbols displayed on the scope overlaying the radar sweep display. A bold line is an Artificial Horizon reference; when the aircraft is in level flight, this line is horizontal and in the center of the display vertically. The small Target Marker Circle shows the relative location of the target the pilot is searching for. If only a portion of this circle is shown, this indicates the target is beyond the range of the present radar display. A bright dot called the Steering Dot shows the azimuth of the Offset Point; if left of center in the display, the plane should be steered to the left to return to the desired course, and if right of the center in the display, the plane should be steered to the right to return to the desired course. When such a steering correction is made, the Steering Dot moves toward the center of the display. When the Steering Dot is in the center of the display, the aircraft is on a collision course with the Offset Point.

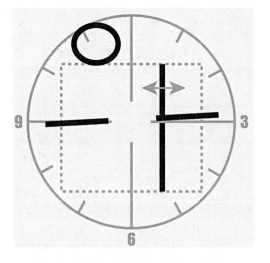

Depiction of MG-10 Fire Control System radar scope symbology during search showing the Target Marker Circle (top left). The Range trace—vertical line with gray arrows—sweeps back and forth across the display, indicating the position of the radar antenna in azimuth. The Horizon Bar shows the aircraft attitude.

A bright white circle with a gap called the Time To Go Circle is displayed beginning when the interceptor is about 2 minutes from the Offset Point. This Time To

Go Circle shrinks toward the center of the display as the interceptor approaches the Offset Point. The radar return is displayed in the center rectangle of the scope. This radar information is presented as a B display, meaning that the distance from the bottom of the scope to a point on the display represents the range, and the distance either left or right of the center of the display represents the azimuth of a return. The Range Trace sweeps back and forth across the display rectangle, indicating the position of the radar antenna in azimuth. The altitude of the target, and hence the altitude the interceptor aircraft should be at, is indicated on a separate panel instrument.

When the interceptor reached the Offset Point, the Time To Go Circle disappeared and the Steering Dot moved out of the center of the display, indicating that the GCI system is sending a different course. This new course is the collision course with the target. It is around this time in the sequence that the target blip in the radar return becomes visible to the pilot. To lock the radar onto this target, the pilot used the left-hand portion of the dual control grip. He moves the left grip sideways to center the range trace horizontally on the target blip, then with a knob on the grip moves the antenna elevation position to obtain the brightest blip possible, then moves the grip forward to align the range mark with the target blip vertically. After the radar is locked onto a target in this manner, the system alters the symbology presented by removing the Target Marker Circle, the Steering Dot moves to a new location, and two concentric circles appear. Centering the Steering Dot places the interceptor on a lead collision course with the point determined by the computer to fire, based on the weapons selected on the Armament Panel. The smaller circle of the two is a Reference Circle and provides a reference for centering the Steering Dot. The larger circle of the two is the Range Circle. The Range Circle remains at a constant diameter until the range to the target is less than 25,000 yards (12.3 nm), after which the diameter of the circle shrinks to indicate the range to target. The position of the gap in the Range Circle indicates the closing speed with the target as read from the tick marks at the edge of the display, at 100 knots per tick mark. By releasing the control stick after lock-on, the pilot engages the Automatic Flight Control System to fly the plane so as to center the Steering Dot.

When the aircraft reaches the point the computer has calculated as the correct point to fire the weapons, the Range Circle and the Reference Circle will have shrunk onto the Steering Dot. The two circles disappear from the scope and a bold X appears in the center of the display, indicating that the weapons have been fired and the pilot should take over control of the aircraft and break off the attack.[27]

Of the Century Series fighters of the mid–Cold War era, four out of the six designs were interceptor aircraft. During this period

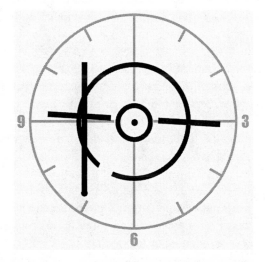

Depiction of MG-10 Fire Control System radar scope symbology during tracking. The Target Marker Circle has disappeared, replaced by the Time To Go Circle and the Steering Dot, shown here locked onto the target.

the technologies and performance of jet aircraft were changing very rapidly. As a result, more than one of the aircraft discussed below were labeled as "interim" because by the time an aircraft and its systems were developed and tested to the point where it could be fielded, newer technology, primarily learned from the recent developments, enabled an even more advanced equipment in a short time.

Aircraft	Range	Speed	Weapons	Fire Control System
F-89H	850 nm	552 knots	6 Falcon, 42 Rockets	E-9
F-89J	850 nm	552 knots	6 Falcon, 2 Genie	MG-12
F-101B	1325 nm	952 knots	2 Falcon, 2 Genie	MG-13
F-102A	1170 nm	720 knots	6 Falcon, 24 Rockets	MG-10
F-104	500 nm	1150 knots	4 Sidewinder, 1 20mm gun	
F-106	1600 nm	1325 knots	4 Falcon, 1 Genie	MA-1

Table of later Cold War fighter interceptor capabilities.

The Northrop F-89H was very little different from the earlier D model, other than incorporating the later fire control system, Hughes MG-12, to allow the primary weapon to become the Falcon guided missiles. Late in its life, the H model was further modified to allow launching the Genie nuclear-tipped rockets intended to destroy a formation of enemy bombers.

The Northrop F-89J was a modified D model brought up to the same functionality as the H model standard, but was given a unique designation to distinguish it from the H's.

The McDonnell Douglas F-101 was initially designed as a long-range penetrator and escort for the Strategic Air Command and was adopted by Tactical Air Command as a fighter-bomber prior to evolving into the B model for the air defense role. All models were powered by two Pratt & Whitney J57 engines to make a total thrust of 30,000 lbs., which was unprecedented for a fighter aircraft at the time.

The Convair F-102 was developed under the project originally called the 1954 Interceptor, and more officially as the MX-1554 project. The F-102 was the first application of the concept of total "weapons system development." This concept meant that the airplane was designed from day one to fully incorporate the radar, fire control system, and supporting equipment. The more common approach at the time called for determining what system components could be back-fit into an existing airframe. This change in design approach came about as a result of the understanding within the ADC leadership that the plane not only had to fly well, but had to complete the desired mission objectives. Such thinking placed greater emphasis on the performance of the system components. Initial flights of the original prototype aircraft indicated that the speed and altitude requirements set by the Air Force would not be met by that design. These requirements were largely set anticipating the availability of high-altitude jet bombers to the Soviet air force. These shortfalls in key requirements led to a significant redesign effort. The primary change in the air frame was redesigning of the fuselage to incorporate the "Area Rule" concept enabling a smooth transition into supersonic flight. The much-improved design went into production somewhat late, and the 1954 Interceptor became operational in 1956. There was a two-seat training version, the TF-102, which had side-by-side seating for an instructor. Most squadrons were set up with one of these planes, which did not have the speed and range performance of the interceptor aircraft.

The Lockheed F-104 single-seat single-engine fighter was powered by the General Electric J79 turbojet. It was ordered by the Air Force as a gap filler until the F-106 was available for the Air Defense Command. The F-104's use by ADC was lessened and shortened due to early problems with the F-104A. However, it should be noted that later models of the plane were long serving with several NATO countries.

The Convair F-106 was originally designated the F-102B, indicating that it was to be the improved version of the F-102. Powered by a single Pratt & Whitney J75 engine, the F-106 was often referred to as the ultimate interceptor. These long-serving fighters were removed from air defense duties in 1988.

The following map depicts the locations of NORAD fighter-interceptor aircraft at the end of 1959. Note that by this time, although the bulk of the aircraft were equipped with the guided missiles discussed above, there were still some squadrons employing the unguided rockets on the F-86 aircraft discussed earlier in this chapter.

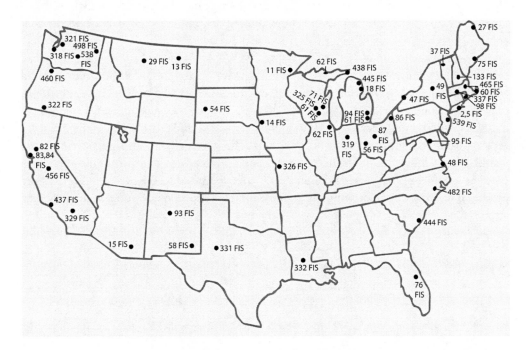

Map of fighter interceptor bases in 1959.[28]

Unmanned Interceptor (missile)

In the later part of World War II, during the period when Germany was launching the unmanned V-1 and V-2 rockets at Allied territories, the U.S. Army asked the Boeing Company to explore the concept of an unmanned air interceptor.

The results of this project went unused until the threat of attack upon the U.S. created a renewed desire for an unmanned defense weapon. In 1950 the Air Force, by then responsible for the air defense, asked the Michigan Air Research Center to join the project. It was the concatenation of the letters for these two organizations that produced the name BOMARC (BO+MARC). This project was in direct competition with the Army's Nike development; due to interservice rivalry, the Air Force wanted its own solution. Congress did fund both programs, at least for a time.

The resulting interceptor vehicle was a bit of a combination airplane and rocket. It was launched vertically with a rocket motor, up to an altitude where it flew as an airplane. The BOMARC missile, following launch from its ground site, was steered along its initial flight path by a sequence of radio commands from a ground-based transmitter. When the missile reached the vicinity of the target, it would take over the guidance function by using its onboard radar to home onto the target.

The BOMARC missile used two different propulsion systems to cover its up to two-hundred-mile range. At launch, a liquid-fueled rocket booster lifted the vehicle off the launch pad. After initial launch, two ramjets joined in to provide sustaining thrust. Upon reaching the cruising altitude of 60,000 feet, it flew like an airplane with wings and tail. At around 10 miles from the target, the homing radar found the target. When the onboard radar locked on to its target, the vehicle would be commanded to enter a dive onto the target aircraft. The 15,000-lb. missile could then attack the inbound aircraft with either conventional high explosive or a nuclear weapon.[29]

With the evolution of the SAGE system, the SAGE computer sent the launch signal, the steering commands, and the final dive command to the BOMARC through a radio channel. The weapons system that incorporated the BOMARC missile included the long-range search and height-finding radars of the SAGE system. When a target was identified, and the weapons controller decided to intercept it with a BOMARC missile, the intercept director would initiate the necessary operations at his console. The launching of the missile would directly follow the controller's hitting the fire button on his console. This would initiate the sending of the steering commands by the computer to the missile. One or more of the radars reporting information into the SAGE Direction Center would then also begin

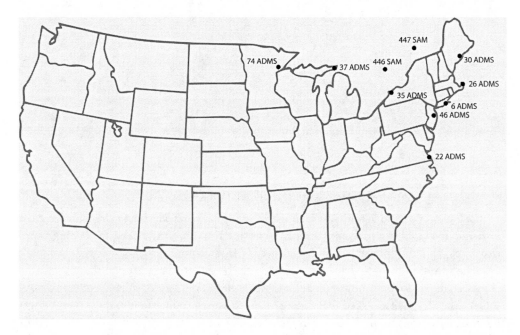

Map of BOMARC interceptor missile sites in the United States.

reporting on the position of the interceptor missile, allowing the computer to track both the interceptor and its target, and to calculate new commands to direct the missile to the point of interception.[30]

There were eight missile squadrons in the U.S. equipped with the two versions of the BOMARC missile that were deployed, beginning in late 1959 at McGuire AFB, New Jersey. The number of missiles in these U.S. squadrons grew to 378. Seven additional sites in the U.S., including four West Coast sites, were constructed but not activated with missiles. As with the Army's Nike missiles, the BOMARC units began to be dismantled as the primary threat of attack shifted to the ICBM. The final BOMARC site ceased operation in 1972.

AIR-INTERCEPT PROCEDURE (MISSILES)

The final aircraft intercept procedure that evolved during the Cold War was done, as the other evolutionary steps before, to increase the lethality of the attack while providing for more safety for the interceptor crew. These improvements were made possible by the evolution of the unguided rocket into the air-to-air guided missile. This evolution was similar to the change from unguided artillery shells into the guided projectile of the surface-to-air missile.

The interceptor aircraft was vectored onto a collision course with the inbound aircraft. This collision course would initially be established by the intercept director assigned to the intercept. The control of the interceptor could be done by radio voice instructions to the air crew, or could be sent by the system to the aircraft's autopilot system. This collision course was intended to bring the interceptor by the most direct route to a common point in the sky where the interceptor would meet the predicted path of the inbound aircraft. The intention of this initial intercept course was to position the interceptor where its radar system could acquire and track the target to the point of launching an initial attack with a forward quadrant weapon.

The reattack intercept differs from the previously used method in that the sequence of events is not only intended to position the interceptor to launch a missile at an intruder, but to do that and position the interceptor to launch a second missile based on results of the first shot. The sequence is to establish a collision course heading to launch the initial missile, then perform a Displacement Turn to preserve or establish a necessary displacement from the intruder's heading line, and then perform a Counter Turn to position the interceptor for the second attack. The result of the initial collision course, for launch of a forward quarter missile, may place the interceptor too close to the intruder's track to permit sufficient maneuvering room to efficiently move the fighter into the proper position for the second rear quarter missile launch. The Displacement Turn allows the pilot to correct for this case.[31]

The weapons, the radar, and other equipment on the interceptor aircraft had to evolve to provide these capabilities. The initial attack from the collision course heading was with a radar guided missile. This type of missile was called Semi-Active Radar Homing or SARH. This means the missile contains a guidance system that steers the missile to a spot from which it receives the radar energy it is designed for. The source of this radar energy is most often the radar in the launching aircraft. The spot the missile receives the radar energy from is intended to be a target aircraft. The interceptor aircraft directs its radar beam onto the target aircraft, which reflects some of that energy back towards the interceptor aircraft, hence the radar receiver on board the missile carried by the aircraft. When the missile signals that it is able to track this reflected target energy, the pilot hears a tone in his headset, and launches the missile. In order for the missile to track such a target long enough for the missile to reach the target, the aircraft must maintain the position of the radar beam on the target, despite maneuvering by both the interceptor and target aircraft. This set of operations is included in the functions of the Fire Control System on the interceptor aircraft, and particularly the radar's ability to track, or keep its beam on, the target, after the pilot initiates this tracking by "locking on" a target.

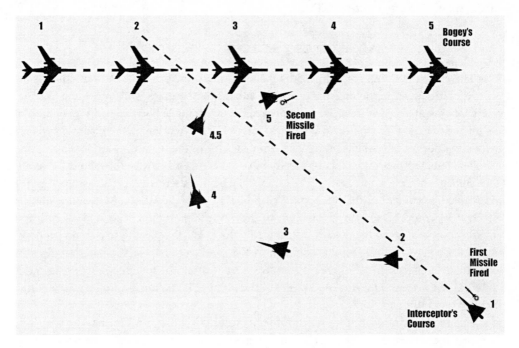

Diagram of sequence of aircraft positions in the Reattack attack used with air-to-air guided missiles.

The second missile launched following the Counter Turn by the interceptor was normally an infrared (IR) guided missile. This missile, an infrared homing type, also has a guidance system that receives IR energy and steers the missile to the spot of the highest intensity of the energy it receives. The source of the IR energy is intended to be the target aircraft. These missiles were initially primarily useful against jet aircraft, whose jet engine exhaust provided an adequate source of IR energy, or heat, for this type of missile to home onto. In order for the missile to receive this source of IR energy, the pilot would maneuver the fighter into the area behind the target aircraft, where an unobstructed view of the target's exhaust existed. This positioning of the interceptor aircraft was most often accomplished using the interceptor's radar. Similar to the radar-guided missile, when the IR missile signals that it is able to track this IR energy, the pilot hears a tone in his headset, different from the tone for the radar-guided missile, and launches the missile. Once the IR missile is launched, the interceptor can turn away immediately, since this missile needs no further input from the aircraft.

This IR energy is electromagnetic energy, at what is called beyond visible range. This means it cannot be seen unaided by the human eye, but is detected by instruments constructed for this purpose, such as the receiver in the IR homing missile. This IR energy is often called thermal energy, because it is a function of the thermal properties of a material. A hotter object, in the heat or thermal sense, provides a more intense reading from the IR-sensitive material of the detector. These sensing instruments, or sensors, are referred to by the terms infrared, IR, thermal, and heat, essentially interchangeably. A missile incorporating this sensing scheme is referred to as an infrared, IR, thermal, or heat-seeker.

With this set of aircraft, there were multiple configurations of weapons. The Genie rocket was a forward quadrant weapon in the F-89J, F-101, F-104, and F-106 fighters. The Genie rocket carried a 1.5 KT nuclear warhead intended to take out a massed formation of enemy bombers with its large blast radius. The Falcon radar-guided missile was a forward quadrant weapon in the F-89H, F-101, F-102, and F-106 fighters. The rear quadrant weapon took the form of the Falcon IR missile in the F-89H, F-89J, F-101, F-102, and F-106 fighters. These Falcon missiles had a range of 5 nm and the Genie rocket a range of 6 miles.

In the 1970s and 1980s, newer aircraft and air-to-air missiles were assigned to the air defense mission, but the techniques and procedures described here were the methods of their application through to the end of the Cold War.

The McDonnell Douglas F-4 Phantom, equipped first with Falcon missiles, then later by the Raytheon Sparrow radar-guided missile and the Raytheon or Ford Aerospace Sidewinder IR-guided missile, was widely used by both the U.S. Air Force and Navy, and by numerous other nations' air services as well. These twin-engine two-seat aircraft replaced the long-serving Century Series fighters (F-101, F-102, and F-106) in the air defense role as the earlier aircraft were retired. Initially, the Sparrow radar-guided missile had a range of 5 nm and the Sidewinder IR-guided missile a range of 3 nm. Both of these missiles had extensive evolution following their mid–1950s debuts, with the missile range being a key parameter for improvement. Both of these missiles are still active today in later variants.

Chapter 3

Defense Against Submarines

The decision to protect the North American continent from the submarine threat carried with it a huge commitment in terms of people, equipment, and techniques. The World War II experience was somewhat limited in that the primary targets at that time were the oceangoing ships, often organized in convoys, needed to supply armies in the fields of Europe and areas of the Pacific. These areas were often limited to the North Atlantic route between Europe and the East Coast Canadian provinces of Newfoundland and the further north Labrador. The Pacific war did not lead to the same need to oppose submarines because Japan had a much smaller submarine force. During the period of the Cold War, once submarines were capable of launching ballistic missiles, all of the ocean areas within range of those missiles from the coastlines were suitable hiding spaces for the threatening boats. As the Soviet Union was large, similar to the U.S., it had seaports on both ends of the Europe-Asia landmass and could have naval presence in both the Atlantic and Pacific, although their ports were some distance from the Atlantic proper.

The combination of commitment and this oceanic geography required the U.S. to provide protection from attack from both oceans, requiring forces sufficient for the dispersal to multiple bases on each coastline, and sufficient fleets for operations in each as well. Additionally, there were the strategically important Hawaiian Islands, which were base and potential target simultaneously.

The number of Soviet submarines would grow continually during the period of the late 1950s. The capability of Soviet submarines would increase up to the 1990s, when many observers believe that the tremendous spending on defense required to maintain parity with the western nations effectively bankrupted the Soviet Union. The peak of their underwater fleet reached 440 diesel-powered boats in the 1950s. Near the end of the Cold War, the Soviets had just fewer than 200 nuclear-powered boats available for operations.[1]

WORLD WAR II EXPERIENCE

A major lesson for the allies of World War I was the benefit of organizing the ocean transport ships hauling supplies to the European war into convoys, so a group of naval vessels could provide protection for a number of transports against attack. This practice was reinstituted even before the U.S. entered World War II.

At the start of World War II, the U.S. had very modest antisubmarine forces. Germany took advantage of this upon the entry of the U.S. into the war and sunk many ships in U.S. waters. This "happy time" for German U-boats was created by the combination of numerous and easy targets with little risk to themselves.

After their occupation of France in June 1940, the Germans moved much of their submarine fleet to bases on the western coast of France, bounded by the Bay of Biscay. This basing significantly reduced the distance the German submarines had to travel to the North Atlantic shipping lanes to attack the eastbound convoys carrying the much-needed war and other material produced by the fields and factories of North America. Additionally, these bases allowed the German submarines to avoid the English Channel, where British forces had the advantage of nearby bases, creating the ability to constantly monitor for U-boat activity. These German bases were only a short distance from the beaches chosen for the massive invasion by the Allies of Fortress Europe in June 1944, making it easier to interfere with the shipping for that invasion. The number of German U-boats would grow to 120 during the war.

Even though the U.S. wanted to focus all resources on the offensive program in Europe to defeat the Nazis, there were a couple of hindrances: the war was a long way off necessitating, a long communication line to Europe; and the relatively advanced German submarine capability would have devastating impact on that shipping. The solution was to devote additional resources to defeating or limiting the effectiveness of the German submarines. Germany's motive in this battle along North Atlantic shipping lanes was twofold: first, destroying the materials being shipped to Europe would lessen the Allies' ability to wage war on Germany; and secondly, those forces allotted by the Allies to the antisubmarine war were not available for any other type of war. Opposing the attacks on Allied shipping altered this equation of supply—more material got through, and the Germans were forced to devote additional emphasis on their submarine forces to achieve reduced results.

World War II submarines could not effectively attack a land target, so primarily only ocean surface targets needed protection. During the war, submarine hunting evolved to include two key and often mutually supporting branches: the sea-based aircraft carrier, with supporting destroyers, a mobile solution; and the land-based long-range aircraft. A sub-branch of the aircraft operations were the coastal patrol bombers that operated near the U.S. coasts. By the end of the war both had evolved the use of sensors and weapons. The primary ship-based sensors were the surface search radar and the sonar. The primary aircraft-based sensor was the radar. Additionally, detections were made by HF-DF, termed Huff Duff, high-frequency direction finding. These radio receivers were primarily placed on land and on surface ships, where reception of a submarine-originated radio transmission by at least two receivers provided a fix on the submarine's location, which the defending aircraft and/or destroyer escort ships could investigate. Supporting these was the airborne searchlight, or Leigh Light, named after the British officer who championed and optimized this needed source of illumination. The searchlight beam was movable in azimuth and somewhat in elevation to enable the air crew to locate and attack a submarine previously detected by radar at night.

Because of the need for vast numbers of naval ships to conduct convoy operations in both the Atlantic and Pacific theaters, the protection of shipping convoys quickly shifted to the destroyer escort (DE) to free the larger and more heavily armed fleet destroyers (DD) for other tasks. These Buckley class DE destroyers were built in large numbers, both to supply the British with additional naval ships, and to provide the U.S. Navy with large numbers of general purpose combat ships.[2] Beginning in late 1942, four of these DE destroyers were typically assigned to accompany an escort aircraft carrier as the escort group for a convoy passing from the northeast U.S. or southeast Canada to Europe, generally England.

The DE destroyer was equipped with a QC sonar and SL search radar sensors. For submarine attack it was equipped with depth charge launchers and torpedoes, in addition to the 3-inch deck guns. This search radar could detect a submarine periscope out to approximately 10 miles. The QC sonar could detect the submerged submarine at about 3600 yards (2 miles).

The escort carrier, the CVE, would be the base for an air group consisting of one squadron of Grumman TBF Avengers and one squadron of Grumman F4F Wildcats. When patrolling for or responding to a sighting of an enemy submarine, one of each type of these single-engine aircraft would operate in concert. The TBF carried the sensing equipment and some combination of antisubmarine weapons: depth charges, bombs, torpedoes, or rockets, as well as machine guns. The Avenger carried the search radar, initially the ASB and later the APS-4, which could detect a submarine from 12 miles to either side of the aircraft or 8 miles ahead, and a crew of three to operate the sensors and radios, as well as the installed machine gun. The F4F, in addition to providing protection of the generally slower Avenger, could use its speed advantage to sprint ahead to a spotted submarine and, via an attack with its machine guns, create confusion or damage to prevent the submarine from using its primary defense: submersion.

In 1944, as the number of carriers, destroyers, and aircraft increased, these Hunter-Killer task forces would continue to accompany a convoy of transports. Other similarly equipped task forces would be assigned to seek out enemy submarines in a particular area of the ocean. The areas selected for these offensive missions were established by the fleet command staff based on a combination of long-range radio direction finding, decoding of enemy radio communications, and other intelligence information.

The long-range radio direction finding (RDF) was accomplished by shore-based receiving stations, of which over 35 were located in England and 4 along the east coast of Canada, to monitor the north and central areas of the Atlantic Ocean. Additional RDF information was gathered as more and more surface ships received the required radio equipment. All of this information was used to dispatch the antisubmarine forces described to the areas of known or suspected enemy submarines as soon as possible after a radio transmission was identified.

The use of long-range land-based maritime patrol aircraft evolved during this time. For the U.S., this role was filled by the Consolidated B-24, or the Navy designation of it, the PB4Y-1. This four-radial-engine airplane was intended and began life as a long-range heavy bomber for the U.S. Army Air Forces. Due to the need to provide protection for shipping over the long distance from North America to various European ports, the range and payload capabilities of this aircraft drew it into the antisubmarine mission. The B-24 was initially used in this role by the British Royal Air Force Coastal Command. Shortly after this the U.S. Navy adopted it for the same role. This aircraft had a cruise speed of 173 knots and a patrol endurance of 4.5 hours at a distance of 1000 nautical miles from its shore-based runway. Its total endurance could be as much as 14 hours, which is why it was selected for this mission.

The submarines of the day were really what are termed submersible. This is a ship with the means of operating beneath the surface for a period of time. During the time submersed, the vessel operates its propeller by electric motor, which receives its power from batteries. The batteries in turn are charged by generators powered by diesel engines on the ship. These diesel engines require air, and so the submarine must operate with access to the air for the diesel while it is charging the batteries. The limitation on the amount of time that the submarine can operate submersed is the amount of battery life on a single charge. All of these facts dictated that the basic operation of this type of submarine is to operate on the surface using diesel engine power both for the propeller and the generators to charge the storage batteries as the normal operation. The submarine operated beneath the surface, using the stored battery power when needed, to hide from an adversary. Additionally, when operating submerged with no access to the outside air, the crew would consume the fresh air inside the submarine over a period of time. The crew's discomfort when the fresh air was consumed often forced the submarine to seek the surface regardless of remaining battery power. This fundamental ability and limitation of the submarine of World War II created and allowed the tactics used on both sides of the conflict. Before war's end, German submarines had begun to incorporate the snorkel, a tube that could provide access to air at the surface, while most of the submarine was under water. This allowed operation of diesel engines with very little exposed above the surface that could be detected. The snorkel complicated the detection of the submarine by the Allied forces.

The German submarine crews, while on the surface, were ever vigilant for the approach of Allied patrol aircraft, which they could often detect by a radar signal sensing receiver. This receiver, known as Metox, could detect the ASV Mk I or II radar at up to 30 miles. The primary defense against the approaching aircraft was to submerge immediately. The problem for the aircraft crew was how to attack the boat which had submerged a short time previously. As the conflict evolved, weapons for this purpose were developed to fill this need. The depth bomb was augmented by rearward firing rockets, which were released by the aircraft after passing overhead the point of submersion by the submarine, and later the homing torpedo which used sonar pulses to seek the submarine despite maneuvering by the submarine.

Evolution in the sensors used by Allied antisubmarine aircraft grew to include the first sonobuoy and the first magnetic anomaly detector (MAD) sensors. The sonobuoy was a small expendable buoy dropped from an aircraft where a submerged submarine was thought to be hiding. These sonobuoys were made up of the hydrophone sensor and a radio transmitter. The hydrophone, or underwater microphone, transmitted to the overhead aircraft by the radio whatever sounds emanated from the water around it. By listening sequentially to several buoys dropped in a geometrical pattern, the operator onboard the aircraft would determine where the submarine was moving by the relative sound levels received and re-transmitted by each of the buoys. The magnetic airborne detector, or MAD, sensed a change in the earth's magnetic field, caused by a large mass of ferrous material, such as a ship or submarine. This sensing was only usable for a distance of a few hundred feet, and required the sensing aircraft to pass directly over the object at a very low altitude to have any value at all.

Each of these innovations in either detection or attack was brought about by an adjustment in tactics by the Germans to nullify the previous threat to their operation. For example, when the German Naval staff realized that their submarines were being attacked following radar detection by an Allied patrol aircraft, their initial response was to operate their submarines on the surface only at night, when the Allied aircraft could not attack because they could not see the submarine. The Allied response was the use of the Leigh Light to illuminate the surfaced submarine following radar detection. This created the situation whereby the Allies could attack the surfaced submarine following radar detection in either day or night conditions. This situation required the Germans to devise the means of learning when radar detection was possible, for which they began to use the Metox radar warning receiver. The accompanying change in submarine tactics was to immediately submerge upon a warning of radar signal reception and remain below the surface for a sufficient period that the patrol aircraft would depart the area.

A further innovation by the patrol aircraft organization was the coordinated use of the patrol aircraft and their respective radars. Named for the intended effect of not allowing any Nazi submarines to enter what would become the Normandy Invasion beach areas of the French coastal regions, sort of a reverse bottling up, these Cork Patrols took advantage of the knowledge learned that the German submarines could not charge their batteries, and as a result could not operate submerged, if they could not remain at the surface to run their diesel engines for at least thirty minutes at a time. The Cork Patrol established a rectangular pattern, with the width of the rectangle being

Diagram of a Cork Patrol pattern.

twice the detection range of the surface radar, and the length of the pattern being dependent on the width, such that the entire perimeter could be flown in 30 minutes.[3]

The theory is that the submarine would detect the presence of radar signals while on the surface and dive to avoid detection. The reason for the submarine being on the surface in the first place is either to communicate, employ its own sensors (antiaircraft), or run the diesel engine to recharge its batteries. By forcing the submarine to dive to avoid detection, assuming no detection, the antisubmarine aircraft prevents the batteries of the submarine from being recharged, resulting, after a short while, in a submarine with no battery power, hence no way to operate submerged. If the submarine is in fact detected, attack would follow.

These Cork Patrol tactics were created for and applied to

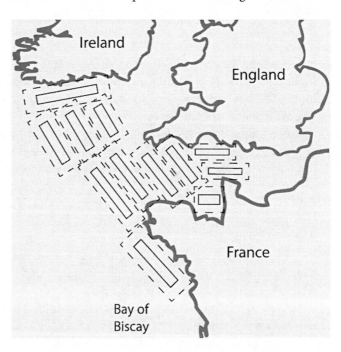

Map of Cork Patrol areas around French coasts.

the coast of France, a major area of Allied shipping activity, where prevention of success by the German submarine arm was crucial to success for the Allies.

In the above map depiction, the solid rectangles inside show the track of the searching aircraft and the dashed rectangular boxes depict the extent of the radar coverage during the patrol.

The successful blunting of the German submarine arm as the European war entered what would be the final, but long, phase of conflict yielded multiple benefits to the Allies. As the Allied offensive moved across France, Germany lost these convenient bases from which to launch the formerly crippling raids into the North Atlantic. As the buildup of liberating Allied forces continued, the new construction and outfitting of even more capable submarines within Germany was overwhelmed by the massive bombing that the continued buildup of Allied material and forces enabled.

By war's end, the key technologies and methods for finding, tracking, and attacking hostile submarines for the following fifty years had all been invented or identified and used to some degree of success by the Allied nations. These methods and equipment resulted in containing, if not defeating, the substantial capability of Germany's subsurface navy. These key technologies were:

1. Long-range direction-finding receivers to locate the source of a boat utilizing its radio transmitter.
2. Surface search radar to detect boats, masts, and snorkels that were exposed above the sea surface.
3. Ship-based active sonar to detect, localize, and track a submerged boat.
4. ECM receivers to detect the presence of a boat utilizing its own air search radar.
5. Passive sonar to localize and track a submerged boat once detected.
6. Magnetic anomaly instruments to localize the submerged boat and setup for attack.
7. Bombs and rockets for attack of a surfaced submarine.
8. Homing torpedoes for attack of a submerged submarine.

The key methods in the use of these technologies were the following:

a. A screen of antisubmarine ships to defend the important assets, either military or transport.
b. Vectoring of surface ships into the area so that their ability to remain and provide sonar coverage allowed the shore-based aircraft to return to their bases for fuel.
c. Periodic aerial overflight of a submarine to keep the snorkel-equipped diesel boat submerged.
d. Strategically located airfields to allow aircraft within range of key transit zones.
e. Carrier-based aircraft for mobility.

In the postwar era, a principal goal and tool of the U.S. antisubmarine effort was to have in place a system to track the location of submarines on a worldwide basis. This is largely necessitated by those vast ocean distances surrounding the North American continent. There is a tremendous amount of water for a potential submarine adversary to hide in adjacent to this land mass.

Confounding the task of keeping track of a relatively small number of hostile submarines in this vast set of oceans are the thousands of cargo ships moving to and from the

U.S. ports of commerce at any time. These ships, having no incentive for stealth, create a huge din of background noise to the main mode of sensing the movement of submarines: sound or acoustic signals.

In the next fifty years of the Cold War, it was primarily these technologies and methods that were refined and expanded upon as the searching to be done was expanded greatly to include the waters of much of the Northern Hemisphere. During the initial period of Cold War, the antisubmarine task was that of protecting the oceangoing convoys that would be needed to supply the next European war, following the presumed Soviet attack on Western Europe.

Most often the process of locating a submarine begins with prior information from some other source than the searching vehicle, such as submerged offshore sensors, information from ships or submarines, or an older piece of information. This last possibility takes the form of knowing where a submarine was one, two, or more days ago and its direction at that time. The information from any of these sources is used to develop a prediction of where the submarine might be now, and that estimate establishes the boundaries for a search.

Finding Submarines at the End of World War II

SURFACE SHIP METHODS

We will now illustrate the operations and tactics of the surface ships that were assigned to the antisubmarine problem.[4] These ships are normally called destroyers and are general purpose ships equipped with sensors and weapons specifically to locate and attack submarines as well as defend against air attack. Their role and title as destroyer escort is that of defending a fleet from attack by submarines or aircraft. The name destroyer is short for the original name associated with the task of defending the most important ships of the fleet: Torpedo Boat Destroyer. This name came about when the new threat to a naval fleet was a torpedo boat. The torpedo boat would maneuver inside the defensive perimeter established around the fleet and launch its torpedo at the battleship near the center. To defend against this new threat, a smaller and more maneuverable ship was developed to seek out and attack the torpedo boat, hence the name Torpedo Boat Destroyer.

The antisubmarine surface ship, the destroyer, had the following sensors available to perform search and identification of hostile submarines:

a. The Mark I eyeball, certainly aided by optical aids such as binoculars.
b. Radar. Two sets normally installed: one for surface search and one for air search.
c. Sonar.
d. High Frequency / Direction Finder (HF/DF)

The range of these sensors is affected by the ship's being on the sea surface. Both visual and radar observation are limited by the line of sight from the ship, ultimately by curvature of the earth. For example, at a height of 100 feet, the line of sight to the horizon is approximately 12 miles. The ship of the World War II time frame often had a visual observation station as high from the water as possible, but still limited to about 65 feet. Similarly the radar antenna could be located no higher than this. There were most often two types of radar on the destroyers: a surface search radar primarily for detecting ships on the surface, and an air search radar primarily for detecting approaching aircraft. The sonar range is that of active sonar, meaning that the signal originated at the ship and had to make the round trip to any target. This range did not exceed 2500 yards for the equipment of the day in any circumstances.

The High Frequency direction-finding sensor was particularly important against the German submarine fleet in the Atlantic Ocean. The Germans were meticulous in their use of radio communications both to receive information from command headquarters and to keep the headquarters updated on the status of each individual submarine. This meant that the submarine commander would approach the surface and initiate radio communication periodically. Both being near the surface and the radio transmission were exposures that the Allied navies frequently were able to take advantage of. Conversely, the HF-DF capability was little used in the Pacific region because the Japanese submarines largely operated in radio silence.

The procedures that follow are for ships assigned primarily to convoy protection. Several antisubmarine ships would be assigned to accompany a group of cargo vessels, say, moving from east-coast ports of North America to England. The same procedures would apply if the ships being escorted were a group of naval vessels. The outstanding advantage of the surface ship in antisubmarine warfare is the ability to remain on station for long periods of time. Additionally, the surface ships offered somewhat more variety in the weapons employed against submarines: two types of depth bomb, the homing torpedo, and the naval gun. The two types of depth bomb were the British-developed Hedgehog, which consisted of a set of charges which was "thrown" ahead of the ship; and bombs that were launched to the side and rear of the ship. These weapons would result in a pattern of explosions that could be placed by maneuvering the ship to the required distance from a submarine to be attacked.

The ships of the antisubmarine screen were stationed around the ships being escorted in generally two patterns: loose screen or tight screen. In either case each ship uses its sonar to echo-range the ocean around the ship. For the first case, the loose screen, the objective of the echo-search plan is to generate a high coverage rate. To obtain the best coverage rate, the range of the maximum expected echo range is used. The coverage rate increases with the speed of the ship up to the point where the water noise from the increased speed reduces the maximum echo range. The ships are stationed about the escorted vessels to search out the maximum area with their sonar beams. The high coverage rate reduces the likelihood that a submarine can penetrate this

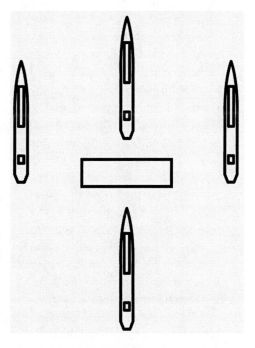

Diagram of Loose Screen ship formation around the escorted convoy.

protective screen without detection. The Standard Echo Search Plan, as described below, is used to move the sonar beam about the area.

For the second case, the tight screen, the objective of the echo search plan is maximum probability of detection. This case is only possible when enough ships can be devoted to the screening so that the spacing between adjacent ships is 1.5 times the assured echo range of the sonar. In order for a submarine to penetrate such a screen, it must risk detection. To accomplish this, the sonar is used with the keying interval for the assured sonar range. The 5° steps of the Standard Echo Search Plan, as described below, are used to move the sonar beam about the area.[5]

If contact is made with the sonar search, the immediate objective is to classify the contact as one of: submarine, non-submarine, or doubtful.

The operations discussed and illustrated below would be performed in

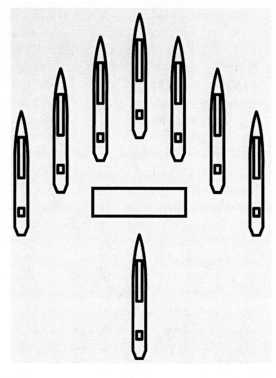

Diagram of Tight Screen ship formation around the escorted convoy.

response to identifying through one of the sensing modes, or by a report of another vessel or aircraft, the possible presence of a submarine. Any of these sensing methods, including an

event such as a ship receiving damage or having been fired upon, would establish the last known point of the submarine. This point would become the initial point for the search that would follow. If an antisubmarine ship is able to reach this last known point of the submarine within ten minutes' time, the following Operation Observant is performed by the first ships to reach the area. If no ship is able to reach this initial point within ten minutes, then Operation Observant is skipped and the Retiring Search Plan is initiated directly.

The pictured Operation Observant would be performed with the initial ship to arrive establishing the Initial Course and marking the Ini-

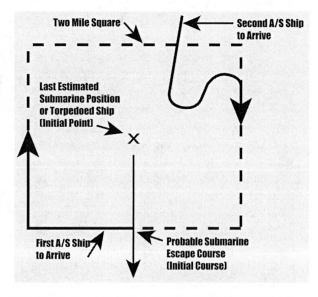

Diagram of Operation Observant pattern about the Initial Point.

tial Point with a marker. A second or third ship to arrive joined the pattern established by the initial ship, on the opposite side of the square. This pattern is only executed a single time. When the entire perimeter of the two-mile square has been patrolled without establishing contact on the submarine, this pattern is discontinued and the Retiring Search Plan pattern initiated. If contact is established, this may lead to another two-mile-square pattern being established about a different Initial Point.[6]

The Retiring Search Plan is a method of employing destroyers to systematically search the expanding area in which a submarine could be located following the loss of contact after the event that led to the initiation of the search. This area grows quickly as the submarine is assumed to travel at 5 knots for the first half hour, then at 3 knots for the next two and a half hours, and then at 2.5 knots after that. Since the ship, and particularly a formation of ships, has difficulty performing a spiral pattern, a sequence of legs with either 90° or 45° turns is completed to move the ships about an increasing radius from the point of the last sighting. A Master Table containing a variety of pre-computed search plans was common to all ships' crews.

In determining which plan to execute, the time until the first ship reaches the last known point of the submarine is determined. This time is called "Time Late." The second parameter affecting the selection of the plan to be done is the assured sonar range, and the last number needed is the number of antisubmarine ships that will participate in the search operation. These values contribute to the distance between individual ships, distance between parallel legs of the pattern, and the length of the sides of the pattern, in order to grow the search area at a rate that will contain the escaping submarine. These values are used to enter the Master Table to determine the plan number appropriate for the situation at hand.

The Time Late value is adjusted if it is thought that the submarine is using a snorkel in its escape. While operating with a snorkel, the submarine is able to run its diesel engines, which results in travel at a faster rate than without. To compensate, the search pattern must grow at a faster rate to ensure containment of the submarine. The following table makes the necessary adjustment.

Actual Elapsed Time	*Time Late*
0:11–0:45	0:11–0:45
0:46–1:30	0:46–1:30
1:31–2:00	1:31–2:30
2:01–2:30	2:31–3:30
2:31–3:00	3:31–4:30
3:01-----	4:31–5:30

Table of Actual Elapsed Times and Time Late values.

If it was thought that the submarine is not using a snorkel, then the actual elapsed time is used as the Time Late to enter the Master Table.

The speed selected by the senior commander on the scene would be the highest speed the ships are capable of, since the higher speed increases the probability of obtaining contact. This speed is selected keeping in mind fuel availability, sea state, and echo-ranging conditions. Additionally, the speed required of a particular plan has been set so that no speed changes are necessary for making the turns. For all of the plans, the distance between adjacent ships during the search would be 1.75 times the assured sonar range.

While proceeding to the last reported position of the submarine, the senior officer directing the search determines the specific plan to be executed, the speed of execution, the distance between ships, the course for the first leg of the plan, whether the plan is to be performed clockwise or counterclockwise, and the expected time the plan is to begin. This information is passed to all other ships to be participating in the plan. The ships proceed to the point where the search is to begin at the maximum practical speed and establish their position relative to the commander's ship, which is positioned in the middle of the group of ships searching. The search begins when the lead ship passes the last reported point of the submarine. This time is signaled to all the ships and they all reset their time-keeping as this time corresponds to 0:00 of the plan.

The search plan itself is a table of time values and turn angles to be accomplished in order to sweep through an ever-expanding area. Each plan refers to a diagram that shows the sequence of legs connected by the sequence of turns. The diagram is a general depiction of the path achieved by the plan, but it is the time points in the table to execute the turns which is the primary information for executing the plan.

There is more than one type of turn the plan may call for. There are 90° turns from line-of-bearing into columns, 90° turns from column into line-of-bearing, 90° echelon turns, and 45° echelon turns. The 45° echelon turn is illustrated in the following figure, because it is the most common turn called for in the plan tables.

In the turn illustrated here, after the first ship has turned, the second ship turns when the first ship is alongside the rear of the ship, the third ship turns when the second ship is alongside the rear of the ship, etc. This would place the ship that has just completed turning directly alongside of the ships which have already turned. As a result of this timing, each ship crosses the path and behind the next ship to turn in the sequence.[7]

During the retiring search operation, radar search would be continuous. If possible, an aircraft provides air coverage to reduce the possibility that the submarine will surface beyond the visible range of the searching ships. If an aircraft is not available, and there are more than three ships, one ship could be detached to perform a radar search. This detached ship moves to the opposite side of the search pattern from the remainder of the searching ships.

When contact is made on a suspected submarine, one or more ships are detached to investigate more fully. During this time the remainder of the searching ships continue per the plan. If the contact is determined not to be a submarine, the detached investigating ships rejoin the main search group to continue the search. The retiring search plan is continued either until

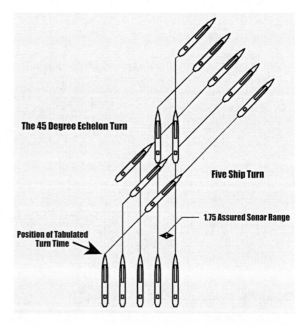

Diagram of a five-ship 45° echelon turn to the right.

contact on the target is made or the search has continued for approximately twenty hours without gaining contact. After this period of searching, the likelihood of gaining contact is small and not worth the continued effort.

Below is the Master Table. This is the map from the various parameters about the search to be initiated to the individual plans and their time tables. The number of ships to be in the search, the assured sonar range, and the time late value that the searching ships will reach the starting point are the entry parameters. This table contains the number of the search plan to be used for that combination of conditions.

The Master Table of search patterns for surface ships[8]

No. of Ships	Assured Sonar Range in Yards	Number of Plan to be Used					
		"Time Late"					
		0:11–0:45	0:46–1:30	1:31–2:30	2:31–3:30	3:31–4:30	4:31–5:30
1	-	1	2	3	4	5	6
2	Less than 2000	7	8	9	10	11	12
	2000–2500	13	14	15	16	17	18
	2600–3500*	19	20	21	22	23	24
	Greater than 3500*	25	26	27	28	29	30
3	Less than 1300	7	8	9	10	11	12
	1300–1800	13	14	15	16	17	18
	1900–2500	19	20	21	22	23	24
	2600–2900*	25	26	26	27	29	29
	Greater than 2900*	31	32	33	34	35	36
4	Less than 1000	7	8	9	10	11	12
	1000–1300	13	14	15	16	17	18
	1400–1700	19	20	21	22	23	24
	1800–2100	25	26	27	28	29	30
	2200–2500	31	32	33	34	35	36
	Greater than 2500*	37	38	39	40	41	42
5	Less than 1100	13	14	15	16	17	18
	1100–1300	19	20	21	22	23	24
	1400–1700	25	26	27	28	29	30
	1800–2000	31	32	33	34	35	36
	2100–2500	37	38	39	40	41	42
6	Less than 900	13	14	15	16	17	18
	900–1100	19	20	21	22	23	24
	1200–1400	25	26	27	28	29	30
	1500–1700	31	32	33	34	35	36
	1800–2100	37	38	39	40	41	42

*Sonar available in World War II was not expected to provide range of greater than 2500 yards.

Below is illustrated one search plan and its associated diagram. In the procedures at the end of World War II there were 42 plans defined and 13 associated diagrams. Here the process will be illustrated with a single plan and diagram. Each plan contains columns of turn time values for execution from 10 to 20 knots, so between the various speeds of searching and the spread of time late values there are a large number of possibilities of search plans to be performed. By having these variances set out in the plans, many ships of slightly different capabilities and many crews were able to have a common set of procedures to be performed as needed.

The following table illustrates the courses, turns, and navigation leg timing for this one example of a retiring search plan. This example is Plan Number 4, which is used with the Diagram 2 shown here.

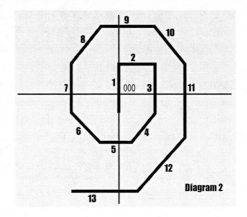

Diagram 2 referred to in the Master Table.

Fast Speeds Diagram Number 2

Leg No.	Change of Course	Search Speed in Knots					
		15	16	17	18	19	20
1	00	0 00	0 00	0 00	0 00	0 00	0 00
2	90	0 43	0 39	0 36	0 34	0 32	0 30
3	90	1 34	1 27	1 20	1 14	1 09	1 05
4	45	2 45	2 31	2 19	2 09	2 00	1 52
5	45	3 32	3 14	2 58	2 45	2 33	2 23
6	45	4 23	4 00	3 40	3 23	3 09	2 56
7	45	5 19	4 50	4 25	4 05	3 47	3 31
8	45	6 20	5 44	5 14	4 49	4 27	4 08
9	45	7 26	6 43	6 06	5 36	5 10	4 48
10	45	8 37	7 46	7 02	6 27	5 56	5 30
11	45	9 55	8 54	8 03	7 21	6 46	6 15
12	45	11 19	10 08	9 08	8 20	7 38	7 03
13	45	12 51	11 28	10 18	9 22	8 34	7 53
14	45	14 31	12 54	11 33	10 30	9 34	8 47
15	45	16 19	14 27	12 54	11 42	10 38	9 45
16	45	18 17	16 08	14 22	12 59	11 47	10 46
17	45	20 24	17 58	15 50	14 22	13 00	11 51
18	45		19 56	17 37	15 51	14 18	13 00
19	45		22 04	19 26	17 26	15 41	14 14
20	45			21 23	19 08	17 10	15 32
21	45				20 58	18 44	16 56
22	45					20 26	18 25
23	45						19 59
24	45						21 40

Table of legs, turns, and turn times for Plan Number 4 referred to in the Master Table. The diagram and plan table shown above are one example of the numerous scenarios the plans allow for. Plan 4 appears in the Master Table for the case in which one ship is to be used in the search and the Time Late value for that ship's arrival at the last known point of contact for the submarine is between two hours thirty-one minutes (2:31) and three hours thirty minutes (3:30).

Let's take an example to illustrate the use of these various tables to determine the search to be done in an example case. Let's say the ship to perform the search is 36 nautical miles distant when it is notified that it must proceed to and initiate a search at Point X, and that the commander chooses to approach the search area at 16 knots and to perform the search at 16 knots. This results in a 2.25-hour transit time, or 2 hours 15 minutes. Since it is believed that the submarine would be using its snorkel to exit the area, the Snorkel Time Late Adjustment table is used to determine that this Actual Elapsed Time leads to using the Time Late value of 2:31–3:30, when consulting the Master Table to determine which search plan is to be used. The Master Table for one ship in the search and this Time Late value indicates Plan 4 as the search plan to be used. The Plan 4 time and turn table indicates that Diagram 2 is the reference diagram.

This search pattern grows the slightly elongated octagon slightly faster in the direction of the initial course. After, in this case, just over 22 hours, the size of the octagon is approximately 65 miles long in the direction of the initial course, and approximately 58 miles wide. If a German Type IX submarine had in fact been traveling in the direction of the initial course at time of submergence and employed its maximum submerged speed as long as possible, it would have reached a point 57 nm from the center of the search. At roughly the same time, the ship adhering to the search pattern would at the farthest search line from the center of the search and crossing the initial course line at approximately 38 nm from the center of the search. This might appear that the ship has wasted its time searching, since in this illustration the submarine would have escaped. However, if the submarine truly used all its battery endurance traveling to the limit of its submerged range, it would then be out of battery power and either forced to surface, exposing itself to radar or visual detection, or forced to remain beneath the surface with no power. Even in this example where the submarine may have avoided detection throughout the search, the search and avoidance have left the submarine very vulnerable. The expanding coverage of the retiring search pattern by the ship also covered the large area where the submarine could well be if it did not continue in a straight line for the full period of its endurance.

Even this simple example of establishing a case of a submarine being sighted and specifying the parameters of the searching force is an illustration of a simple war game. What if the ship had been closer to the point of observation? What if the ship had searched at a higher speed? What if the submarine had made turns during its escape path? The first two of these possibilities would have the search run a slightly different path by entering the tables with different initial parameters. The possibilities are close enough to endless, and that is another reason for the tables and the diagrams: to allow the local commander to select from among a set of standard plans, all being started with the same entry parameters, yet each altering the resulting search pattern enough to account for the differences in time and speed of the specific case.

ECHO SEARCH PLANS

The following illustrates the use of the ship's sonar during any search operation, including the search plan illustrated above. Illustrated in the diagram below is the Standard Echo Search Plan with a 5° step between each transmission, or ping, of the sonar. Shown by the numbers on the diagram is the sequence of the angles set sequentially in the search. The time between pings, or keying interval, is set to allow for the sonar pulse to travel out to an intended range and return from a target at that range before the sonar transmits again. The speed that sound travels in water varies with water conditions, but an average value for this speed is 5000 ft/sec. To make a 2000-yard round trip, a sonar pulse would travel this distance in 2.4 seconds, and it would make a 3000-yard round trip in 3.6 seconds.

The Standard Echo Search plan is done by use of the automatic keying operation of the sonar. This operation creates the sonar pulse and moves the sonar projector at the rate selected to generate the sequence of pulses at the various angles depicted in the following diagram. The 5° angle between successive pulses is the most efficient for this ping-train-listen sequence of the search. If this angle is larger, the efficiency decreases. If ships are close enough that an adjacent ship causes interference with the sonar, a different angle from the one shown in the diagram would be selected to reduce the size of the search sector for that ship so that the sonar operates without interference.[9]

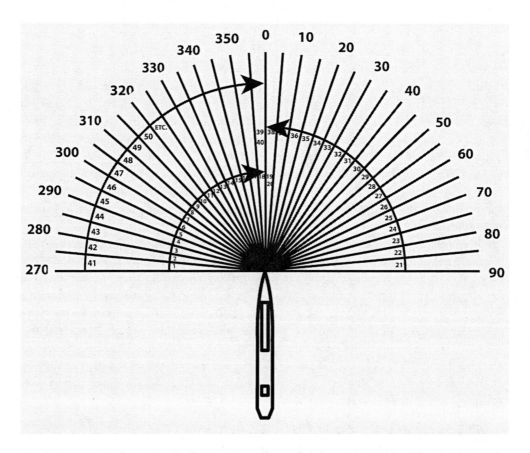

Diagram showing the sequence of azimuth positions for the sonar pings of the Standard Echo Search plan.

USE OF SONAR EQUIPMENT

Sonar operators are selected for a variety of attributes, including aural acuity, reaction and coordination, focus, intelligence, and ability to communicate. Tone perception is essential. In addition to skills taught in operating the sonar equipment, the sonar operators are also required to be versed in the navigation-related topics of true and relative bearings, relative motion between ship and target, and the degree of change in range and bearing that may be expected from a submarine during attack by the ship. To achieve the levels of ability desired, these skills are honed through frequent practice in echo ranging and listening on friendly ships.

Each type of sonar had switches to control the timing of the pulse transmitted. The values to be set are the pulse length and the time between successive pulses. The sonars either had a switch to select from a "short" range pulse or a "long" range pulse, or a switch to alter the pulse length. If the equipment has the Short-Long Scale Switch, during calibration of the sonar, these two selections were set to between 30 and 40 yards and between 150 and 170 yards, respectively. These pulse length distances correspond to transmit time periods of 22 milliseconds and 100 milliseconds, respectively. The long outgoing signal is required to obtain long-range echoes; the long signal is not required at shorter ranges. The long signal was needed when the maximum sonar range is more than 2000 yards. The following table shows the resulting time between pulses for the common selectable values. This time is sufficient for the sonar pulse to travel out to a target at the distance, and return to the sonar receiver. When tracking a target, it is common to use the smallest signal interval time possible because this will increase the number of pulses transmitted over time, and provide the operator with more return signals to interpret. The range scale would be selected large enough to allow for maneuvers of the target about the current range. The automatic keying intervals available from representative equipment and their settings are summarized in the following table.

Description	*Signal Interval Switch*	*Short Scale-Long Scale Switch*	*Interval*	
			(Yards)	*(Seconds)*
Short Range	1 M		1000	1.25
Medium Range	2 M		2000	2.50
Intermediate Range	3 M		3000	3.75
Intermediate Range	4 M		4000	5.00*
Long Range	5 M		5000	6.25*
Extreme Range	10 M		10000	12.50*

* Not used during search.

Table of common sonar switch settings to search out to various ranges from the ship.

The sonar was capable of either Automatic or Manual keying. Automatic keying was used to implement the prescribed echo search plan. The plan assumes that the sonar transmissions are maintained at a definite interval so that the intended area of the search is completely swept by the beam of successive sonar pulses. Manual keying was used in other situations when a continuous string of pulses was not desirable. To investigate a situation, the operator might send a distinctive signal in order to identify weak or doubtful echoes.

During coordinated operations the operator might want to prevent interference with the outgoing signals (based on automatic keying) of a nearby friendly ship. A short signal would be used during the final stages of the attack or whenever the maximum echo-range is less than 2000 yards.

Sonar Range Recorder. In order to partially automate the submarine tracking and attack process, the Sonar Range Recorder was adopted. This device used a moving paper being marked by a stylus pen to record the history of the sonar range over time. In order to do this, the Range Recorder would take over the keying operation of the sonar. After triggering the sonar transmission, the recorder began moving the stylus from the left edge of the paper, the point associated with a time of zero. The stylus continued moving at a constant rate until a signal returned from the transmitted pulse caused a mark on the paper. For each sweep of the stylus, after either a signal return had been detected, or the pen reached the right edge of the paper, the stylus was quickly moved back to the left edge of the paper, and the next sonar pulse initiated. The scaling of the plot was such that the operator could think, and report, distances on the plot in either yards or time. A mechanical attachment called the plotter bar, after setup by the operator, allowed the range rate to be read directly from the scale provided.

The history of the range to the target was used to predict when the range would reach the correct value for the launching of weapons. Although this prediction was of greatest value if both ship and submarine maintained their course and speed, the range trace also showed any change in the relative motion between the vehicles; if the ship was not changing direction or speed, any change observed on the trace must be the result of changes by the submarine.

The Range Recorder also took over the sending of signals to launch the individual depth charges to accomplish the pattern of charges for an attack. The Range Recorder had settings to insert the delay times particular to a depth charge attack accounting for time of flight and sinking time to the selected depth. Properly set up, it was said that the precision in the launch commands was better than humans could accomplish with the stopwatch.

The recorded plot from an attack sequence could be reviewed to identify mistakes in any of the crew's actions. This feature could obviously be used in training of new crewmen. The plot was also useful if the returned signal was weak or affected by other conditions of the water such as fish or debris. The trained operator could interpret the recording when these conditions were present and pick out the likely returns from the target submarine.

Listening for submarine hydrophone effects during the interval between outgoing signals while echo ranging would provide the advantage of evaluating the submarine or non-submarine nature of a contact and may afford an indication of the submarine's speed. Training for operators included opportunities to familiarize themselves with the sound of submarine and torpedo propellers, so that operators could recognize torpedo noise and be alert to report the bearing when detected. In addition to target noises, propeller noises from the ASW ship could be heard on some bearings and had to be distinguished from the target. Similarly, enemy submarines could be equipped with echo-ranging equipment, and sonar operators would be alert to detect and report echo-ranging signals that could not be identified as those of the sonar screen, i.e., friendly ships.

Determination of Maximum Echo Range. The maximum echo range on targets at the surface would be determined by echo-ranging a large surface ship underway. The range would be determined by opening out on the surface ship until contact is lost or by closing in until contact is gained. To determine the variation of maximum echo range with depth, the bathythermograph would be used.

USE OF DEPTH CHARGES

Similar to the standard tables for search patterns, there were tables for the launching of depth charges. The tables contained a variety of plans to account for the variables of the type of depth charge to be used, the type of impact charge that would launch the weapon away from the ship, and the number and location of launchers on the ship. The officers would select the plan that covered the arrangement of the ship and the equipment to be used on a particular attack. This selection would not vary greatly on an individual ship, as the location of launchers was fixed.

Characteristics of Depth Charges

Types and Sinking Rates

Mark 6	300 lbs. TNT	Total weight 420 lbs.
Mark 7	600 lbs. TNT	Total weight 768 lbs.
Mark 8	270 lbs. TNT	Total weight 520 lbs.
Mark 9 or Mark 9–1	200 lbs. TNT	Total weight 320 lbs.
Mark 9–2, 3 and Mark 14	195 lbs. TNT	Total weight 340 lbs.

Table of depth charges available for anti-submarine operations in World War II.

The Master Plan, on following page, was consulted to determine the pattern most applicable to the ship's armament. The procedure is illustrated with the 13-charge pattern commonly used by the DE class of ships in either the A or B pattern of the Master Plan. For each of the patterns in the Master Plan table, only five firing orders are needed. A pair of side throwers would be fired with the signal to drop from the stern tracks, and is illustrated in the sample firing orders. Shifts of the side-thrown charges up to 40 yards along the line of advance are acceptable as long as no charge is within 100 feet of any other charge in the pattern.

Projectors

	Depth Charge Mark 6		Depth Charge Mark 8		Depth Charges Mark 9 and Mods. and Mark 14	
Impulse Charge						
Number	*Range*	*Time of Flight*	*Range*	*Time of Flight*	*Range*	*Time of Flight*
1	50 yds.	3.0 secs.	43 yds.	2.8 secs.	60 yds.	3.4 secs.
2	75 yds.	3.8 secs	63 yds.	3.5 secs	90 yds.	4.2 secs
3	120 yds.	4.7 secs	90 yds.	4.2 secs	150 yds.	5.1 secs

Table of Depth Charge Projector Impulses used with the various Depth Charges during World War II.

The sinking times of depth charges vary with the type of wake produced by the specific ship, the ship's speed, leafing effects when falling through the water, angle that the charge strikes the water, and other unique effects of the situation.

Sinking Time in Seconds

Depth of Water in Feet	50	100	150	200	250	300	350	400	450	500	550	600	650	700	750	800	850	900	950	1000
Mark 6	10.0	16.0	22.1	28.2	34.3	40.4	46.5	52.6	58.7	64.8	70.9	77.0	83.1	89.2	95.3	101.4	107.5	113.6	119.7	125.8
Mark 6 with Mk 7 arbor attached	5.5	10.8	16.1	21.4	26.7	32.0	37.3	42.6	47.9	53.2	58.5	63.8	69.1	74.4	79.7	85.0	90.3	95.6	100.9	106.2
Mark 7	10.0	16.0	21.5	27.0	32.5	38.0	43.5	49.0	54.5	60.0	65.5	71.0	76.5	82.0	87.5	93.0	98.5	104.0	109.5	115.0
Mark 8 thrown from projectors	4.5	8.8	13.1	17.4	21.7	26.0	30.3	34.6	38.9	43.2	47.5	51.8	56.1	60.4	64.7	69.0	73.3	77.6	81.9	86.2
Mark 9 or Mk 9–1	4.1	7.6	11.1	14.6	18.1	21.6	25.1	28.6	32.1	35.6	39.1	42.6	46.1	49.6	53.1	56.6	60.1	63.6	67.1	70.6
Mark 9 or Mk 9–1 thrown from projectors	3.8	7.3	10.8	14.3	17.8	21.3	24.8	28.3	31.8	35.3	38.8	42.3	45.8	49.3	52.8	56.3	59.8	63.3	66.8	70.3
Mark 9–2,3 and Mark 14	3.1	5.3	7.5	9.7	11.9	14.1	16.3	18.5	20.7	22.9	25.1	27.3	29.5	31.7	33.9	36.1	38.3	40.5	42.7	44.9
Mark 9–2,3 and Mark 14 thrown from projectors	2.6	4.8	7.0	9.2	11.4	13.6	15.8	18.0	20.2	22.4	24.6	26.8	29.0	31.2	33.4	35.6	37.8	40.0	42.2	44.4

Table of sinking times to reach various depths for the depth charges available in World War II.

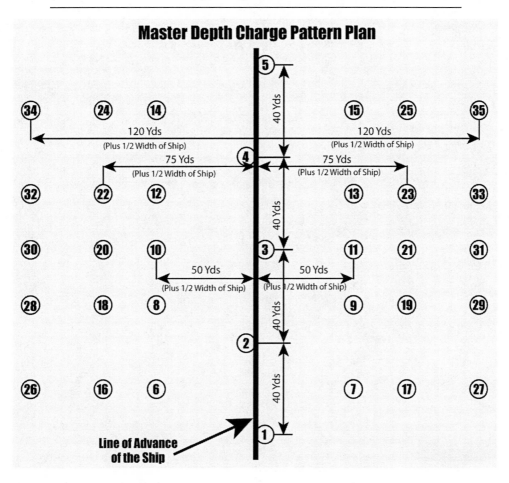

Master Depth Charge Pattern Plan

Line of Advance of the Ship

Diagram showing the resulting patterns achievable by the plans represented in the Master Table.

7 Charge Pattern	Fire	1	2	3	4	5	10	11		
Settings	A	100'	50'	100'	50'	100'	150'	150'		
	B	250'	200'	250'	200'	250'	300'	300'		

7 Charge Pattern	Fire	1	2	3	4	5	20	21		
Settings	C	400'	350'	400'	350'	400'	450'	450'		
	D	550'	500'	550'	500'	550'	600'	600'		

7 Charge Pattern	Fire	1	2	3	4	5	30	31		
Settings	E	650'	700'	650'	700'	650'	800'	800'		
	F	850'	900'	850'	900'	850'	1000'	1000'		

9 Charge Pattern	Fire	1	2	3	4	5	8	9	12	13
Settings	A	100'	50'	100'	50'	100'	150'	150'	150'	150'
	B	250'	200'	250'	200'	250'	300'	300'	300'	300'

9 Charge Pattern	Fire	1	2	3	4	5	18	19	22	23
Settings	C	400'	350'	400'	350'	400'	450'	450'	450'	450'
D		500'	500'	550'	500'	550'	600'	600'	600'	600'

9 Charge Pattern	Fire	1	2	3	4	5	28	29	32	33
Settings	E	650'	700'	650'	700'	650'	800'	800'	800'	800'
	F	850'	900'	850'	900'	850'	1000'	1000'	1000'	1000'

11 Charge Pattern	Fire	1	2	3	4	5	6	7	8	9	12	13
Settings	A	100'	50'	100'	50'	100'	150'	150'	100'	100'	150'	150'
	B	250'	200'	250'	200'	250'	300'	300'	250'	250'	300'	300'

11 Charge Pattern	Fire	1	2	3	4	5	16	17	18	19	22	23
Settings	C	400'	350'	400'	350'	400'	450'	450'	400'	400'	450'	450'
	D	550'	500'	550'	500'	550'	600'	600'	550'	550'	600'	600'

11 Charge Pattern	Fire	1	2	3	4	5	26	27	28	29	32	33
Settings	E	650'	700'	650'	700'	650'	800'	800'	750'	750'	800'	800'
	F	850'	900'	850'	900'	850'	1000'	1000'	950'	950'	1000'	1000'

13 Charge Pattern	Fire	1	2	3	4	5	6	7	8	9	12	13	14	15
Settings	A	100'	50'	100'	50'	100'	150'	100'	100'	150'	150'	100'	100'	150'
	B	250'	200'	250'	200'	250'	300'	250'	250'	300'	300'	250'	250'	300'

13 Charge Pattern	Fire	1	2	3	4	5	16	17	18	19	22	23	24	25
Settings	C	400'	350'	400'	350'	400'	450'	400'	400'	450'	450'	400'	400'	450'
	D	550'	500'	550'	500'	550'	600'	550'	550'	600'	600'	550'	550'	600'

13 Charge Pattern	Fire	1	2	3	4	5	26	27	28	29	32	33	34	35
Settings	E	650'	700'	650'	700'	650'	800'	800'	750'	750'	750'	750'	800'	800'
	F	850'	900'	850'	900'	850'	1000'	1000'	950'	950'	950'	950'	1000'	1000'

Standard Pattern Settings are Designated by the Letters A, B, C, D, E, and F

Note: #1 impulse charges were to be used on all A and B patterns, #2 impulse charges on all C and D patterns, and #3 impulse charges on all E and F patterns.

Table of the available Depth Charge patterns used in World War II.

The following illustrates the sequence of events and the resulting pattern of charges for a 13-charge pattern from the DE-type ship. This type of ship is equipped with two tracks on the stern, or rear, of the ship and four side throwers on each side of the ship.

Instructions for Firing 13 Charge DE Pattern

Sonar Range Recorder	Distance from First Charge	Time from First Charge at 15 Knots	Time from First Charge at 18 Knots	Order	Stern Tracks	Side Throwers
Fire 1st				"Fire One" One Short Blast	Release 1 DC	Fire Throwers 7&8
Fire 2nd	40 yds.	4.7 secs.	4 secs.	"Fire Two" One Short Blast	Release 1 DC	Fire Throwers 5&6
Fire Center	80 yds.	9.5 secs.	8 secs.	"Fire Center" Two Short Blasts	Release 1 DC	Fire Throwers 3&4

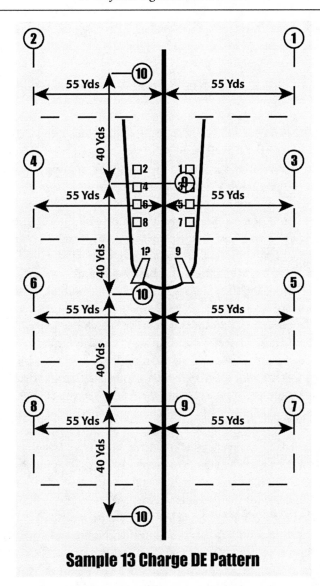

Sample 13 Charge DE Pattern

Diagram of the resulting location of depth charges resulting from the 13 Charge DE Pattern.

Sonar Range Recorder	Distance from First Charge	Time from First Charge at 15 Knots	Time from First Charge at 18 Knots	Order	Stern Tracks	Side Throwers
Fire 4th	120 yds.	14 secs.	12 secs.	"Fire Four" One Short Blast	Release 1 DC	Fire Throwers 1&2
Fire 5th	160 yds.	19 secs.	16 secs.	"Fire Five" One Short Blast	Release 1 DC	

Impulse Charges: All throwers loaded with #1 Impulse Charges.

Table of the depth charge launcher firing sequence for the 13-charge DE pattern.

Observe that this pattern results in a rectangular pattern of charges 160 yards long by 110 yards wide.

Preparation for Attack

When a searching ship detected an unidentified target, either through radar search, visual search, sonar search, radio direction finding, or information outside the ship, this contact would be investigated. This investigation was initiated as quickly as conditions permitted in order to characterize the contact as either threatening or nonthreatening. To accomplish this, one or preferably two ships proceeded toward the contact at the maximum practical speed. During the approach, preparations for attack with guns, torpedoes, or to ram the target were made. Sonar contact would be made with the target as soon as possible. When within range, fire with guns and torpedoes could begin, and again the ship prepared for ramming the submarine if necessary. Should the submarine submerge, the bearing to the swirl was used as a reference to initiate a shallow standard pattern of depth charges. If the swirl was not visible at the time of arrival at the point of submergence, speed would be reduced to allow for operation of the sonar. If sonar contact was not made, then the previously described Operation Observant or a Retiring Search Pattern was initiated, depending on whether a ship could arrive at the point within ten minutes.

In order to prepare for all possibilities surrounding an attack on a submarine, it was vital for a ship approaching for a possible attack to evaluate the movement of the target submarine, by determining the range rate and the rate of bearing change to the target. Similarly, it was necessary to determine the depth of the submarine following submergence so that weapons could be set accordingly to obtain maximum destructive results. In many cases the submarine would limit its dive to between 150 and 250 feet, but the submarine was quite capable of diving to 800 feet when necessary. A rough guideline for the depth to which a submarine might have reached is: the range in yards at the time of loss of contact is approximately the depth of the submarine in feet. For example, if contact is lost at 300 yards, then the submarine could have reached 300 feet in depth.

Attacks by single ships fall into two categories: deliberate and urgent. In the deliberate attack there is no immediate danger to a convoy or other asset, and the attack is made following location and tracking of the submarine. The forward-firing weapons, Hedgehog or Mousetrap, would normally be used before the stern- and side-launched depth charges. The urgent attack is one in which there is immediate danger to another vessel or asset. This attack is made without all of the preparation of the deliberate attack. The urgent attack is made when the submarine is detected, either by sight, radar, or sonar, when at a position from which a torpedo could be launched at the escorted vessels. The stern- and side-launched depth charges are always used in the case of urgent attack, and the attack is repeated as long as the urgency, i.e., the immediate threat to the protected assets, exists. Although the urgent attack could be made without all the sensor indications of the deliberate attack, those systems would be used during the urgent attack, although the attack could not be delayed due to failure of any of these additional sensors. Regardless of acquisition by the various sensors, the urgent attack was made to provide timely protection to the protected assets.[10]

Attacks in which multiple ships, normally two, are present are known as coordinated attacks and put the two ships in the roles of attacking and assisting ships. The primary advantage gained by the two-ship section attack is the increased ability of two ships to maintain contact with the submerged submarine. If the submarine should be able to counter the methods of the attacking ship, the assisting ship can still maintain contact and conduct additional attack if the attacking ship or its equipment became disabled for any reason. If a third ship was available on the scene, it would establish a box search around the scene as a supporting ship while the attacking and assisting ships prosecuted the target.

The first type of coordinated attack is the two-ship attack on a submarine that is not known to have gone deep. The attacking ship is the first ship to make contact with the submarine and would normally make the initial attack. However, if the first ship to make contact is in a poor position to initiate the attack, it might communicate to the assisting ship to become the attacking ship. If the first ship turns over the attack to the assisting ship, the first ship then takes on the role of the assisting ship and communicates any information on the submarine to the attacking ship. The assisting ship moves to a position 1000 to 1500 yards from the submarine. This position is best to enable the assisting ship to maintain contact and report submarine movement and to be ready to mount an attack if directed by the attacking ship. If contact with the submarine is lost by both attacking and assisting ships, Operation Observant would be initiated, as previously described.

The second type of coordinated attack is the two-ship attack on a submarine that is known to have gone deep. The following attack was intended to enable the assisting ship to maintain contact on the submarine and to prevent the submarine from being aware of the approaching attacking ship. This approach improved the likelihood of destruction of the very deep (over 400 feet) submarine. The first ship to make contact took on the role of assisting ship. The assisting ship maneuvered to a position 1500 to 2000 yards from the submarine. A position astern of the submarine assisted in deception of the submarine as to the position of the attacking ship. The assisting ship maintained a plot of the position of the submarine and the attacking ship. The assisting ship used voice radio to direct the attacking ship courses to steer and when to initiate depth charge attacks. The attacking ship moved to a position 1000 yards ahead of the assisting ship between the assisting ship and the submarine. The attacking ship followed the courses directed by the assisting ship and at the slowest speed that would allow the attacking ship to overtake the submarine. The attacking ship did not use sonar during this operation and initiated the depth charge sequence when directed by the assisting ship. After the first charge of the pattern exploded, the fathometer was used to determine the depth of the submarine and a change to the depth setting of the remainder of the charges in the pattern was made as appropriate.[11]

The antisubmarine ship had a sonar attack team, on duty at all times, which conducted these operations. This team was made up of the conning officer, the sonar officer, the recorder operator, the plotter, the sonar operator, and the standby sonar operator. Each of these positions had specific duties during the various phases of an attack on an enemy submarine.

Conning Officer. Directs the maneuvering of the ship, gives orders to launch weapons, and determines the lead angle for approach to achieve a lead collision course on the target.

Sonar Officer. Directs the other members of the attack team, determines the water conditions as they affect sonar operations from the bathythermograph measurements and information from other sources, orders the settings for the sonar transmissions, orders any change in the sonar receiver settings, directs the angular limits of the sonar search, directs the depth charge party and the Recorder Operator of depth charge settings to be used, and directs the Sonar Operator and the Recorder Operator that the recorder is to take over control of the sonar transmission.

Recorder Operator. Sets up the firing corrections on the recorder based on the attack speed and the depth charge settings, determines the firing time from the recorder traces, and reports range, range rate, and target inclination.

Plotter. Maintains a continuous plot of the target and the attacking ship, advises the Sonar Officer of areas to conduct a search if contact is lost.

Sonar Operator. Operates the sonar in an echo search by stepping toward the bow after beginning at the aftermost bearing on that side of the ship, operates the sonar when the order "Investigate Arc from—to—" or "Make Listening Sweep from—to—" with the angles ordered, reports the result, returns to previous search if result is negative, when an echo is received, delimits the target at the after limit first, then delimits forward limit, reports forward cut-on as it changes so Conning Officer can determine lead angle, and reports all contacts, characterized as: submarine, non-sub, or doubtful,

Standby Sonar Operator. Reports ranges from the sonar instruments until transmission is shifted to the recorder, reports Doppler indication, computes and reports the center bearing when requested by the Sonar Officer, is ready to take over operation of the sonar to alternate every ½ hour with the operator, or to relieve the operator if his hearing becomes degraded.[12]

A time-range-bearing plot would be created and used for each attack on a known or suspected submarine. The Plotter would use the mechanical plotting aid, the Anti-Submarine Attack Plotter (ASAP), if available to make this plot. The Dead Reckoning Tracer (DRT), in the Combat Information Center (CIC), would be used to plot all antisubmarine attacks, in addition to the ASAP. If a mechanical plotting aid was not available, a manual plot would still be made to record the relative positions of the ship and the target using the maneuvering board tool. The maneuvering board is a flat surface with range rings and angles drawn permanently, so that a piece of tracing paper overlaid on the board can quickly be marked with reference points to the ship's position and direction. If contact with the submarine was lost, the plotter continued plotting the ship's position and estimated the submarine's position based on its course and speed prior to losing contact. This information was used to direct the ship to search based on the last known course of the submarine. After regaining contact, the submarine's position, along with the plotted positions prior to loss of contact, indicate the travel of the submarine during the outage, or indicate that the contact is probably not the same as what was lost.

Ranges and bearings determined during the approach are applied to the plot. During the initial period of the approach, the submarine's position would be gathered every 20 seconds. Information derived from the plot were the course and speed of the target. Successive positions plotted during the approach and attack were used to refine the estimate of the submarine's position so that the attack could be made. During the attack the approximate time for the weapons to be launched was extrapolated from the plot.

Together this crew used the sonar readings to bring the ship to close proximity of the path of the submarine to make the attack. Generally the focus was to identify the bow end of the submarine, and its motion so that a lead collision course could be established. A lead angle of around 15° ahead of the target would often achieve the interception. The Conning Officer had been taught to develop a feel for adjustments to the lead angle, through practice in exercises, as opposed to applying many specific rules. The sonar was used to "cross" the target periodically so that the center bearing of the target was in mind by the crew. This also served to confirm the bow location of the submarine. If any change in the sub's course was suspected, a turn to the center bearing was made to assess. Constant monitoring and plotting of the sub's course was done to detect as soon as possible any change in the relative motion so that corrections could be made. Changes in the bearing rate and Doppler were the key values which indicated the submarine was altering course. A constant bearing to the target indicated the approach is working during the initial phase, while the bearing moving aft was expected as the two ships got closer.

The following diagram depicts the desired sequence. When the submarine was initially detected, the ship turned to the bearing of the target to assess the movement of the submarine. After the direction of the submarine was established, an initial lead angle on the submarine's bow was taken by the destroyer to intercept the submarine's track. At or before reaching 700 yards from the target submarine, the destroyer took the attack heading to cross the submarine's track sufficiently ahead to allow time for the falling and sinking depth charges to reach the submarine's depth.[13]

If everybody had done his job well, the ship would cross in front of the track of the submarine. After launch, the depth charges would travel in the air for about 3 seconds and then under water for 16 seconds to reach a depth of 100 feet. The submarine would enter this water under the position where the ship had been some 19 seconds earlier. The ship would be clear of the area when the charges exploded, having moved almost 160 yards in this time. The lead angle had to place the destroyer far enough ahead of the submarine to allow for the sinking time of the weapons to reach the depth of the submarine. Compared to the example depth of 100 feet, the lead angle would be greater, 20° or more, for a submarine at, say, 300 feet. Ships' officers became adept at estimating the travel of the ship, the submarine, and the weapons in this three-dimensional problem.

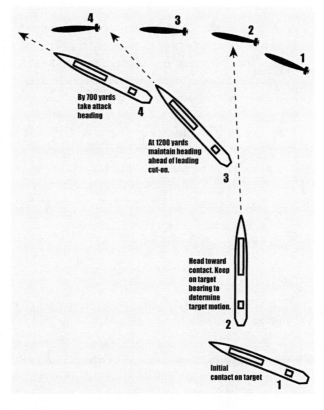

By 700 yards take attack heading 4

At 1200 yards maintain heading ahead of leading cut-on. 3

Head toward contact. Keep on target bearing to determine target motion. 2

Initial contact on target 1

Diagram of the sequence of angles the ship steers upon detection of a hostile submarine.

Following this approach path will place the submarine nearly directly behind the attacking ship after the weapons are launched, in "the baffles" of the ship where the sonar is least effective. A turn toward the last position of the submarine is often made to help the sonar situation. This course must be continued at least far enough to achieve sufficient room for the ship to turn to a proper heading for reattack.

Because the submarine could likely hear the surface ship activities in the closing phases of the attack, it was not often that the destroyer had a straight-line constant course submarine to attack. This led to the development of the homing torpedo weapon, which was intended to follow every maneuver of the submarine, in addition to allowing the attacking vessel to remain further from the hostile submarine. This weapon was adapted for use by both ship and aircraft to attack the submerged submarine.

Aircraft Methods

The methods available for the searching aircraft and crew were heavily dependent on the sensors and other equipment available at the time. By the closing days of World War II, the set of sensors available to the antisubmarine aircrews were:

a. The Mark I eyeball. This could be augmented by optical aids such as binoculars, but the Mark I was the primary sensor available for visual search.

b. Radar. Several radar models were in use with submarine detection ranges from 2 to 25 miles.

c. Sonobuoys. These instruments re-transmitted any detected underwater sounds via radio for reception on the nearby aircraft. The installed acoustic signal processor for these underwater sounds was the Mark I trained human listener. This human could detect presence or absence of sounds of interest, and relative strengths of a common sound on two or more listening sonobuoys.

d. Magnetic Airborne Detector. This instrument detected and displayed the strength of the magnetic field for several hundred feet around the sensor boom on the aircraft. Primarily this sensed the earth's magnetic field, but a large ferrous object, such as a ship or submarine, would cause a disturbance in the normal field, and this disturbance is what the crew is searching for.

The principal reason for an airborne mission of the antisubmarine aircraft and crew was to apply one or more of the above sensors to one of the problems associated with an enemy submarine. Namely:

a. Searching for a submarine following some sort of previous knowledge of the location of the submarine.

b. Establishing a barrier through which a submarine cannot travel without a possibility of detection by the patrolling forces.

To accomplish these missions, the patrolling aircraft, via its crew, would conduct a patrol operation, either searching using visual means or one or more of the electronic sensing systems, or a combination of the sensing capabilities. The objective of the operation was to either identify the location of the enemy submarine or advertise to the submarine presence of the aircraft. With a searching aircraft, passage by the submarine entailed substantial risk of detection and whatever consequences. The search could be accomplished on two levels: by determining either general or precise location, depending on the prior information. Once the location was determined, identification could be followed by attack if appropriate.

AIRBORNE SEARCH

A search by patrol aircraft is implemented by establishing a flight path to cover the desired search area over a period of time. One of the following patterns is used depending on the specific circumstances of the situation. The key dimension of the pattern to be flown is determined in large part by the effective visibility of the sensor to be employed. Effective visibility, E in subsequent paragraphs, is the distance such that if the aircraft passed within this range of the target, the probability of detecting the target with that sensor is high; while if the aircraft passes the target at a greater distance, the probability of detection is relatively low. This second term is the probability of detection or probability of success in the search. The effective visibility is the range where there is a 50 percent probability of detecting a target, in the conditions present, and is affected by altitude, atmospheric conditions, as well as the sea state of the ocean surface. For example, if an area is searched using the parallel tracks method described below, at a track spacing of twice E, the resulting probability of target detection is 50 percent. This may be acceptable for many situations, but does not guarantee success. To increase the likelihood of success, the tracks would need to be placed closer together, increasing the amount of flying time needed, or the number of aircraft to be used, to assure making contact on all possible targets.[14]

The following table shows the effective visibility in nautical miles for airborne search radars in use by the end of World War II, for several search targets.

Object of the Search	ASE	ASB	ASD	APS-3 (ASD-1)	APS-2 (ASG)	APS-15
Snorkel (Sea State 0 to 1)	2	2	3	6	4	9
Raft with corner reflector	2	4	5	9	10	12
Surfaced Submarine	8	8	9	12	22	22
Ship (4000–8000 ton)	22	20	22	30	40	40
Convoy (4 or more ships)	24	22	26	36	50	50

Table of effective visibilities for several targets using the various airborne search radars of World War II.

The basic guidance for radar search is to use the optimum altitude for that radar. For visual search of a surfaced submarine, the airplane flies at the base of the clouds if the cloud cover is more than 5/10, just above the clouds if the cloud cover is less than 5/10, and at 5000 feet if there is no cloud cover. A visual search of a snorkeling submarine, was flown at 1000 feet.

The following table shows the effective visibility in nautical miles for a visual search, again for several types of search target and from several different altitudes.

Object of the Search	*Altitude in Feet*	*Meteorological Visibility in Nautical Miles*							
		3	*5*	*10*	*15*	*20*	*30*	*40*	*50*
Stationary Motor	500	0.7	1.5	3.0	4.0	4.5	5.0	5.5	5.5
Vessel or Combatant	1000	0.7	1.5	3.0	4.5	5.0	6.0	6.5	7.0
	2000			3.5	5.0	6.0	7.5	8.0	8.5

Object of the Search	Altitude in Feet	Meteorological Visibility in Nautical Miles							
		3	5	10	15	20	30	40	50
	3000			3.5	5.5	6.5	8.0	9.0	10.0
	5000				5.5	7.0	9.0	10.5	11.5
Moving Submarine or Small Motor Vessel	500	3.0	4.0	7	8	9	10	11	12
	1000	3.5	5.0	9	10	11	13	14	15
	2000	3.5	5.5	9	11	13	16	17	18
	3000		6.0	10	12	14	17	19	20
	5000			11	14	16	19	21	23
Moving Motor Vessel or Minor Combatant	500	3.5	5.0	11	14	16	19	21	23
	1000	3.5	6.0	12	15	18	23	25	27
	2000	3.5	6.5	13	17	21	26	29	32
	3000		6.5	13	18	22	28	32	35
	5000			13	20	24	31	35	39
	7000				20	25	32	37	42
Moving Major Combatant	500	3.5	6.0	12	16	18	23	25	27
	1000	3.5	6.0	13	17	21	26	29	32
	2000	3.5	6.5	14	19	23	29	34	35
	3000		6.5	14	20	24	32	37	40
	5000			14	21	26	24	40	45
	7000				21	27	36	42	47
	10000					28	37	44	50
Moving Convoy or Task Force	Use the effective visibility for a moving major combatant plus the radius of the convoy or task force.								
One-Man Life Raft	Recommended Altitude in Feet	500	650	900	950	1000	1050	1100	1150
	Effective Visibility	0.25	0.35	0.45	0.45	0.45	0.50	0.50	0.55
Large Raft	Recommended Altitude in Feet	1000	1300	2000	2400	2700	2800	2900	3000
	Effective Visibility	0.6	0.9	1.3	1.5	1.6	1.8	1.9	2.0

Table of effective visibilities for a visual search for several types of search target and from several different altitudes.[15]

When the effective visibility for the planned search has been determined, the search method is selected. The choice of search pattern depends upon the known information about the position, course, and speed of the target of the search. If the number of aircraft available for the search is less than needed to cover the desired area to a high probability of success, it is better to use the available aircraft in a search of the entire area at a lower likelihood of gaining contact than to reduce the search area to only a portion of the planned area.

Aircrews were trained on each of the search patterns described below. This background of common training and procedures allowed for supervising officers to plan out larger searches over time, and assign a particular aircrew to a specific search using one of the following prescribed patterns. Because of the standard method, aircraft from different units including different type of aircraft could be incorporated into a coordinated search.

The parallel sweeps pattern, often called a ladder search, is used for systematic patrol for a target whose position is very uncertain, with course and speed unknown. The distance between tracks of the pattern (S) is nominally 2 * E, to achieve a 50 percent probability of contact.

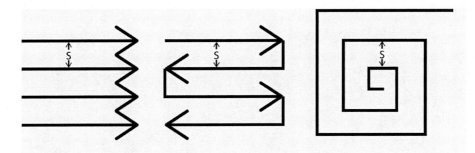

Diagrams of the parallel sweep, ladder, and expanding square search patterns. Dimension is normally twice effective visibility.

The expanding square pattern is used when the target's last position is reported, the course is unknown, and speed is small. The distance between tracks of the pattern (S) is also normally 2 * E.

The aircrew was provided with the following nomograph to determine the sweep spacing. The nomograph is a graphical method of calculating a result from a formula. The scaling of the values on each vertical bar and the spacing between the various vertical bars implements the desired formula. In this case the formula is a relationship that determines the Sweep Spacing based on the values of Effective Visibility and Probability of Contact. In the present day this formula and the two input values would be combined using a pocket calculator or possibly would be included in the programs of a computer on board the aircraft. In the 1940s, those solutions were not available, so the formula was worked out once, and after careful checking, was made into the nomograph you see below. Numerous such nomographs were in the manuals of the aircraft, enabling the crew to calculate values relating to fuel, weight and balance, and the winds affecting the flight path. To use the nomograph, you needed the two input values and a straightedge. As an example, take the effective visibility of 2 nm and determine the Sweep Spacing to achieve a Probability of Contact of 0.5, or 50 percent. You lay the straightedge between the point of 2 nm on the Effective Visibility scale on the left, the point of .50 on the Probability of Contact scale on the right, and read off the value of the point of intersection on the Sweep Spacing scale in the middle, for this example 4 nm.

The trapping square pattern is used when the target's last position is reported, the course is unknown, and speed is estimated. This approach is used for a submarine search if the search can be started within one hour of the last known position report. The trapping square is also called the gambit patrol, whose objective is to induce the submarine to return to the surface and provide an opportunity to the patrolling aircraft to regain contact

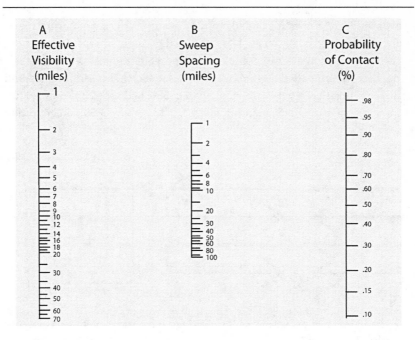

A Effective Visibility (miles)	B Sweep Spacing (miles)	C Probability of Contact (%)

Depiction of the nomograph used to determine Sweep Spacing given Effective Visibility and desired Probability of Contact.[16]

and attack if possible. The radar sensor would be used continuously during this gambit patrol.

The dimension of the trapping square is determined by first estimating the Center Position Error, for the target, and the Distance the object has moved from the reported position. The center position error (P) is determined by adding the expected error in the last known position report of the target and the search aircraft navigation error, which is often 5 percent of the distance from the aircraft's base. The distance the object has moved from the previous position (D) is determined by multiplying the surface speed of the object by the time delay, which is the time in hours from the time of the contact until the start of the search. For sub-merged submarines of World War II, this is taken as 10 knots

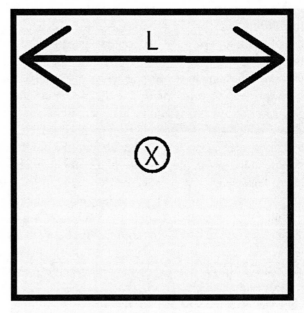

Diagram of the Trapping Square search pattern. Dimension L is four times the center position error plus two times the distance the object has moved.

for the first two hours and 7 knots after that time. The dimension (L) of the trapping square is then four times the center position error + two times the distance the object has moved.

$$L = (4 * P) + (2 * D)$$

For a submerged submarine as target, the length of the trapping square is increased by 20 nautical miles every three hours without additional contact. This pattern is followed for up to 12 hours, after which the area of likely escape is searched.

For a submarine with snorkel, the above approach for the submerged submarine is followed, except that after 8 hours of patrol with no new contact, the length of the leg is expanded to 80 miles to account for the increased speed of a snorkeling submarine over and above what speed is expected from a submerged submarine.

For a surfaced submarine as target, the length of the trapping square could be set according to the speed capability of the searching aircraft. The following table contains the length of the square legs and the time to remain in this pattern for several ranges of speed.

Aircraft Speed	Length of Legs	Maximum Time	Length After 4 Hours	Length After 8 Hours	Length After 9 Hours
125–150	20 miles	4 hours	45 miles	65 miles	75 miles
150–175	25 miles	4 hours	45 miles	65 miles	75 miles
175–225	30 miles	6 hours	45 miles	65 miles	75 miles

Table of Trapping Square legs and time to search for ranges of aircraft speeds.[17]

Aircraft would fly a diagonal of the square pattern to remark the center after two circuits of the pattern, to maintain a watch on the point of last contact. With these expansions due to lack of contact, if possible additional aircraft would be added to the patrol area to increase the likelihood of contact given the increasing area being covered by the longer legs. These area expansions are to make the square large enough to account for possible movement of the submarine, while the addition of an aircraft in the center area is to maintain coverage of the areas previously searched. If contact was not achieved by 12 hours, an additional aircraft would be added to fly a 25-mile length pattern in the center of the larger square. After 18 hours with no contact or other indication that the submarine has remained within the square, this square pattern search would be discontinued, and a sweep of areas where it was determined the submarine could have moved would be made.

When knowledge of the location of a submarine is not known, but it is considered likely that a submarine must pass through an area of ocean transiting either to or from one location to reach some objective, such as a home port, vicinity of the fleet, or some other destination, the antisubmarine barrier patrol may be used to deny passage of the surfaced submarine through a particular channel or area. Due to the normal possibility that a submarine will travel submerged, the probability of making contact with the submarine during that time is decreased, if the search is made by radar or visual means. It was assumed that submarines will travel surfaced if able to do so, and for part of every day they must remain on the surface to recharge their batteries. As mentioned previously, the submarine of World War II preferred to travel on the surface if able, because of the increased speed available from the diesel engines and the greater comfort for the crew due to access to fresh air. These ideas are all taken advantage of in the planning of the barrier air patrol.

The barrier patrol was set up to try to catch the enemy submarine on the surface, while transiting the channel over which the barrier patrol was established. This requires the patrolling aircraft to be launched so as to be in the barrier pattern during the time period that the submarine was anticipated to be surfaced in the area. It is important that the barrier patrol be maintained long enough to cover the entire range of time the expected submarine could enter the area. The factors affecting the establishment of the flight parameters to cover a specified channel are submarine speed, aircraft speed, channel width, aircraft endurance, available aircraft, visibility, and the acceptable probability of contact on the submarine.

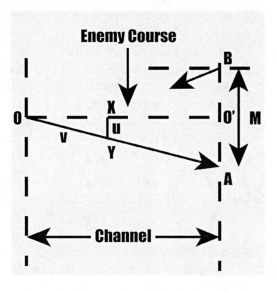

Diagram of the Barrier Patrol search pattern.

The two vertical lines represent the lateral limits of the channel. The starting point for the patrol flight is point O. The submarine is moving along the channel at speed u, whereas the aircraft searches along the line O-A at speed v. The course of the aircraft is set so that the speed of the aircraft along the channel, in the direction the submarine is expected to be traveling, is the same as the speed of the submarine along the channel, u. This allows determining the delta angle from the line directly across the channel O-O.' This angle can be calculated by use of the right triangle O-X-Y. For planning, a surfaced submarine was assumed to be 18 knots for the first hour and 12 knots after that, nearly twice the speed of the submerged operation. Since they did not have pocket calculators to perform this calculation, they would have had a table of pre-computed values for the available aircraft and submarines of the day. Taking a mid-speed value for the various classes of aircraft illustrated previously allows one to estimate a lead angle for each group, for the expected submarine speeds.

Aircraft Speed	Lead Angle @ 18 Knots	Upsweep @ 18 Knots	Lead Angle @ 12 Knots	Upsweep @ 12 Knots
125–150	9°	0.9 * S	6°	0.9 * S
150–175	7.5°	0.9 * S	5°	0.9 * S
175–225	6°	0.9 * S	4°	0.9 * S

Table of Barrier Patrol lead angles given expected submarine speeds for ranges of aircraft speeds.[18]

When the aircraft reaches the far side of the channel, point A in the above diagram, it turns up the channel to begin another sweep across the channel using the same delta angle to account for the expected movement of the submarine. This is called the Upsweep distance, where S is the Sweep Spacing as determined by the same method as used for the Parallel Sweeps search pattern above. If the channel to be patrolled is narrow, the time required to complete one sweep across the channel is less than the time to the next departing flight, to achieve the probability of contact that is built into the Sweep Spacing S. If the channel is wide, the time to complete one sweep across is more than the time until the next aircraft should depart. For all cases the time between departing flights is given by the formula:

$$T = (2 * S) / u$$

These search and barrier patterns as described would provide an organized and standard method of searching for nearly anything by an aircraft. The calculations discussed would have been completed prior to the flight; likely each aircraft squadron would predetermine values to be used with their aircraft, knowing what type of enemy submarine they were expecting. It is when that search locates a submarine that the following sections would apply specifically to the problem of what to do about the submarine once detected.

Following an attack or a sighting of a submarine, the following general steps were completed:

1. As soon as possible after a sighting is made on a submarine, a radio report of the contact is transmitted.

2. The rate of approach to the target of a radar contact was accurately determined so that if the radar contact disappeared prior to a visual sighting, an expected position could be calculated and an approach by dead reckoning made to that point.

Following sighting or attack of a submarine that subsequently submerges, the point of submergence is marked. If the equipment for tracking an underwater submarine is not available, radio sonobuoys or magnetic airborne detector, the gambit search tactic is initiated to reestablish contact with the submarine by inducing it to return to the surface. After an attack, the aircraft would monitor the area for damage to the submarine for 15 minutes before initiating the gambit tactic to regain contact. The Trapping Square with continuous gambit was the most productive and economical. When calculating positions for possible locations of submarines, speeds of 18 knots for the first hour and 12 knots after the first hour for a submarine traveling on the surface would be used. For a submerged submarine, 6 knots for the first ten minutes and 3 knots after that would be used. Note: These speeds are different from those presented in the Trapping Square section. This is to make the Trapping Square large enough to contain a wide variety of movements.

USE OF RADAR

The configuration of airborne search radar in use at the end of World War II evolved out of several types available earlier. With these early types, primarily detection of only a surfaced submarine was possible. As the submarines themselves evolved, detecting only the snorkel above the surface was needed, and this was achieved by the newer radars with the increased frequency of the later designs. The following table shows the antenna configuration of the different types used during the war.

Radar	Beam Coverage	Comments	Coverage Diagram	Search Sweep Width (nm)
ASE	Two 30° lobes forward, two 15° lobes to the sides. No Sweep.	This is a U.S. adaptation of the British ASV Mk II.		16

ASB	Two 40° lobes. Operator controlled from forward to 90° either side, in 8° steps.		16
ASD	5° beam swept through 140°.		17
APS-3	5° beam swept through 160°.		23
APS-2	9° beam swept through 360°.		44
APS-15	3° beam swept through 360°, or selected sector scan.		44

Table of the antenna coverage patterns for the airborne search radars available during World War II.[19]

What became standard out of this evolution was a rotating radar antenna, normally contained in a radome slung beneath the airplane, which rotated to cover 360° in azimuth, but could be limited to an operator-defined sector. This was presented to the operator on a PPI scope display, which depicted the radar beam as a rotating sweep line at the antenna angle and any returned signal at a distance out from the scope center representing the range of that reflection.

The higher the aircraft altitude, the farther out the radar could find a target. The radar signal traveled in a straight line-of-sight path from the aircraft to the horizon, the point where the curvature of the earth makes an object at greater distance invisible. For example, at 1000 feet, the radar horizon is 35 nm; 3000 feet, 70 nm. For the transmitter powers available at the time, the recommended altitude for submarine search was 1000 feet, since this provided 35 nm range, greater than detection range at that time, and left the aircraft low enough to transition for an attack on a detected submarine.

USE OF SONOBUOYS

The radio sonobuoy was created to allow the antisubmarine forces to search for and detect a submerged submarine. The sonobuoy is an air, or otherwise, deployed acoustic sensor that was invented during World War II. The first air-droppable sonobuoys, the AN/CRT-1, were packaged in a three-foot-long cylinder about 5 inches in diameter. This size came to be known as "A" size. This included a half-watt radio transmitter that could be tuned to one of six FM frequencies. The radio transmitted sounds detected by a nickel magnetostriction hydrophone, or underwater microphone, which sank to a depth of 24 feet, for broadband listening. The CRT-1A, ordered in greater quantity, provided 6 color-coded radio frequencies. It was designed to transmit by radio the underwater sounds it receives from the surrounding water. These transmitted sounds from the sonobuoy would normally be received by nearby ships or aircraft. The main uses of these sonobuoys were[20]:

a. Localizing the position of a submerged submarine to a small area so that it could be attacked if need be.

b. Determination of the damage done to a submarine in a prior attack.

c. Investigation of the area of a disappearing radar contact.

d. Investigation of the area of magnetic airborne detection indications.

e. Investigation of the area where debris from a potentially damaged submarine was found.

f. Determining the location of potential survivors of prior attack on a submarine.

Most often multiple sonobuoys were placed in the water, in a pattern, around the area where the last known observation of the submarine was made. The standard pattern of five sonobuoys is the most common, as illustrated below. This last known position of the suspected submarine could be the point where it was observed to submerge, the point of the last observed radar return, the point where magnetic anomaly indications were detected, or the point where debris or an oil slick was found on the surface.

The aircrew is trained to maintain the orientation of the plane to the buoys by the use of sea markers and to maintain an accurate and up-to-date position plot. Normally, the entire pattern of sonobuoys is deployed in the water when the situation calls for sonobuoys so that the multiple buoys can enable the comparative listening approach. It is important that the spacing and orientation of the buoys match the pattern diagrams, illustrated below, as closely as possible because comparative listening is based on the buoys being equal distances apart. The distance between the separate buoys of the pattern determines the probability of detection by a buoy just as the sweep spacing using visual sensing or radar sensing determines the probability in the previously described search methods for an object on the surface. The gambit patrol when aided by a pattern of sonobuoys is useful when there are no other forces in the area to assist the aircraft, and the aircraft is equipped to conduct an attack when a submarine is located. The sonobuoy is capable of detecting the shift from electric motor use to diesel by a submerged submarine. The listening crew member had been trained in the differences in sounds, including the increased RPM noises of the diesel engines, leading to identification of the time when the submarine surfaces.

The following methods were employed to establish a set of sonobuoys in the water around a point where something was to be investigated. Notice the similarity in that all methods result in a pattern of five buoys, equally spaced around the central point of interest. The distance between buoys varies based on the type of information that led to the investigation, but all patterns result in placing a border around the area to contain a submarine if it exists. For example, if the information leading to the investigation was a lost contact based on visual detection, the aircraft was likely closer to the contact than if the lost contact had been a radar contact, and so the containment boundary would be smaller. If the lost contact had been a radar indication, the aircraft was likely farther away, i.e., the aircraft previously had had a radar indication that had not resulted in a visual identification, and so the containment boundary is larger. In each case, the operator aboard the aircraft would sequentially listen to each of the individual radio channels to hear what could be heard from each sonobuoy location.

The following steps were for laying a sonobuoy pattern at the point where a submarine had submerged following previous contact.[21]

1. Fly to the point of submergence and note the heading. 1500 yards prior to reaching the point of submergence, drop the first sonobuoy.

2. Continue on course and drop the second sonobuoy at the point of submergence.

3. Continue on course and drop the third sonobuoy 1500 yards after the point of submergence.

4. Commence a 270° turn to the left. Continue the turn and straighten out on a heading 090° from the original heading. 1500 yards prior to reaching the point of submergence, the location of buoy #2, drop the fourth sonobuoy.

5. Continue on this heading and drop the fifth sonobuoy 1500 yards after the point of submergence.

The adjacent diagram illustrates this pattern. Note that for this pattern the Buoy Spacing is 1500 yards.

The adjacent actions are for laying a sonobuoy pattern at the point of a disappearing radar contact in conjunction with performing a Trapping Square Search Gambit.

1. Drop a five-sonobuoy pattern as in the previous procedure, except use 5000 yards instead of the 1500 yards of the previous procedure.

2. After laying the pattern of sonobuoys around the suspected point of the submarine, fly to perform a normal Trapping Square Search Gambit at a distance of 2 times the visual range of a submarine, then fly the square pattern with each side equal to four times the visual range.

A variation of this procedure was used to investigate almost any indication of possible submarine presence such as oil slicks, air bubbles, or debris on the water. The intention is to get a sensor in the water as quickly as possible, and then base subsequent searching on what is learned from that buoy. If a submarine is not heard on the first sonobuoy, the pattern is discontinued and the crew returns to the previous mission.

If the area where a submarine was thought to be was larger and the preceding methods did not apply, a larger sonobuoy pattern could be established to localize a submarine in an area. Acceptable results from the large sonobuoy pattern would be expected if the submarine was known to be in a limited area, the sea is calm, and the submarine is moving at a speed of 2 knots at least part of the time before the sonobuoys run out of battery life. The guidelines for the pattern were:

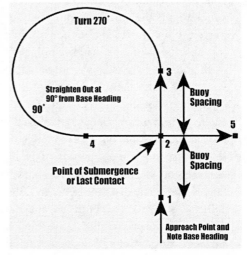

Diagram of the sonobuoy pattern applied at the point of last observation of a suspected submarine.

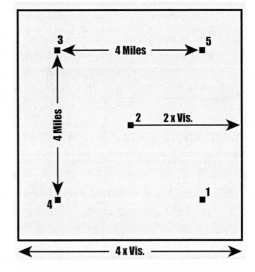

Diagram of the sonobuoy pattern applied at the point of last observation in conjunction with the Trapping Square pattern.[22]

a. The pattern is sufficiently large enough to ensure that the submarine is contained within the boundary.

b. The pattern is either a square or rectangle of equally spaced buoys; the number and probability of contact is determined by the table below.

c. There needed to be one aircraft to monitor the sonobuoys and perform visual and radar search for each six sonobuoys of the pattern.

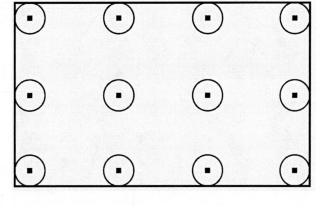

Diagram of the sonobuoy pattern applied to search an area for a suspected submarine.

When an indication of the submarine is heard on a particular frequency, each aircraft would fly over the buoy of that frequency that it is monitoring at low altitude (200–300 feet) to determine which buoy is active. When the active buoy was identified, the normal procedure to develop the contact would be performed.

For a Submarine Search Unit Area of 100 Square Miles
(For a submarine traveling at 2 knots at periscope depth)

Number of Buoys	Probability of Contact
12	.16
18	.24
24	.32
36	.40
42	.47
48	.52

Table of the Probability of Contact for a given number of sonobuoys in a rectangular search pattern.[23]

The above table shows the probability of contact for a pattern with the number of buoys provided per unit area of 100 square miles. If a different area of A square miles were used, the number of buoys for a particular probability is multiplied by A / 100. For example, if sonobuoys were to be used in a 200-square-mile area, then the probability of contact would be:

$$12 * 200 / 100 = 24 \text{ buoys for probability .16}$$
$$18 * 200 / 100 = 36 \text{ buoys to achieve probability .24}$$

When submarine contact was achieved with a basic sonobuoy pattern, the submarine may be followed, or tracked, by extension of the pattern so as to determine the course and speed of the submarine. With this information a position at which to attack the submarine could be determined.

Example 1—Through a Vertex—There is good indication of a submarine on buoy 2, then on buoy 1, but not good indication on either 3, 4, or 5.

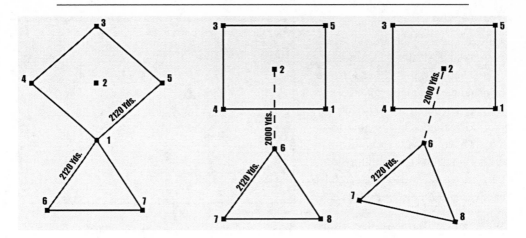

Diagram of location of additional sonobuoys to track the movement of a submerged submarine.[24]

Example 2—Through Center of a Side—There is good indication of a submarine on buoy 2, then equally good indication on 1 and 4.

Example 3—Through Off Center of a Side—There is good indication of a submarine on buoy 2, then on 1 and 4, but much louder on 4.

These examples of using the differing levels of sound from the buoys of a pattern illustrate the concept of comparative listening. The operator compares the level and quality of the sounds he hears in the headset from the various buoys, and judges the relative position of the source to the buoys.

When a buoy, previously silent, began transmitting or when the intensity of the transmission increased, it indicated that:

a. The submarine is approaching the buoy, or
b. The submarine has speeded up

In the case where it has increased speed, if antisubmarine ships are within the submarine's listening range, this may indicate that the submarine has decided upon evasive or offensive action. A submerged submarine could likely hear a 15-knot DD ship at 15 miles in a calm sea. The submarine could likely hear surface vessels at longer range than the submarine itself could be heard on the sonobuoy transmission.

The table below shows ranges of sonobuoy detection on submarines in yards and nautical miles (nm), with the sonobuoys of World War II. Average sound conditions were assumed, and the actual ranges may vary from ¼ to 4 times the ranges shown. The maximum range at which a DE ship can be heard on the sonobuoys under these average conditions is included for comparison.

Speed—knots	*Range—yards [nm]*			
	Sea State Calm	Sea State Smooth	Sea State Slight-Moderate	Sea State Rough
Sub @ Periscope Depth				
2	1150 [.57]	800 [.36]	635 [.31]	500 [.25]
3	2000 [.99]	1300 [.64]	1100 [.54]	900 [.44]

Speed—knots	Range—yards [nm]			
	Sea State Calm	Sea State Smooth	Sea State Slight-Moderate	Sea State Rough
4	3600 [1.7]	2300 [1.1]	2000 [.99]	1600 [.79]
5	5200 [2.6]	4000 [2.0]	3100 [1.5]	2600 [1.3]
6	8900 [4.4]	6050 [3.0]	5080 [2.5]	4080 [2.0]
7	10400 [5.1]	9000 [4.4]	8000 [4.0]	6800 [3.4]
Sub @ 150 feet				
2	795 [.39]	500 [.25]	400 [.20]	305 [.15]
3	1150 [.57]	800 [.40]	635 [.31]	500 [.25]
4	2000 [.99]	1300 [.64]	1100 [.54]	900 [.44]
5	3080 [1.5]	2080 [1.0]	1800 [.89]	1400 [.69]
6	5040 [2.5]	3400 [1.7]	2950 [1.5]	2150 [1.1]
7	8000 [4.0]	5800 [2.9]	4800 [2.4]	3975 [2.0]
Sub @ 250 feet				
2	675 [.33]	420 [.21]	330 [.16]	280 [.14]
3	990 [.49]	610 [.30]	500 [.25]	400 [.20]
4	1200 [.59]	990 [.49]	790 [.39]	610 [.30]
5	2150 [1.1]	1600 [.79]	1150 [.57]	1025 [.51]
6	3600 [1.78]	2300 [1.1]	2000 [.99]	1600 [.79]
7	5200 [2.6]	4000 [2.0]	3100 [1.5]	2600 [1.3]
DE Ship				
4	3080 [1.5]	2080 [1.0]	1800 [.89]	1400 [.69]
6	4100 [2.0]	3025 [1.5]	2150 [1.1]	2020 [1.0]
8	6025 [3.0]	4100 [2.0]	3400 [1.7]	3010 [1.5]
10	9900 [4.9]	6975 [3.4]	5900 [2.9]	4800 [2.4]
12	11000 [5.4]	9900 [4.9]	8050 [4.0]	6800 [3.4]
14	14000 [6.9]	11000 [5.4]	10500 [5.2]	10000 [4.9]

Table of submarine detection ranges for World War II sonobuoys for different conditions.[25]

This table clearly shows that the sound of the submerged submarine increased dramatically with speed, so that it could be detected at a much greater range by the sonobuoy at the increased speed.

In the preceding paragraphs, the distances between sonobuoys is illustrated in yards, the common scale and terminology for the time. For comparison, 1500 yards is 0.85 miles, 3000 yards is 1.7 miles, and 5000 yards is 2.84 miles.

USE OF MAGNETIC AIRBORNE DETECTOR

The Magnetic Airborne Detector (MAD) is a set of equipment on the aircraft to detect a disturbance in the earth's magnetic field, such as what would be caused by a ship or submarine. The MAD is used in antisubmarine operations to confirm the presence and position of a submerged submarine. The MAD detector may be used in conjunction with the previously described Barrier Patrols to detect the passing of a submerged submarine, or the Parallel Sweeps or Expanding Square search patterns may be used to investigate suspicious conditions or material on the surface. The range of the MAD sensor is very small, a few hundred feet, so the aircraft must be flown at the lowest practical altitude for its use to be effective. As a result, searching with this sensor was confined to a small area.

A submarine was located with the MAD sensor through a sequence of trapping and tracking procedures. Trapping is the procedure to establish initial contact on a submarine with the sensor. Tracking is the process of maintaining contact to determine the movement of the submerged submarine after contact is established.

When a submarine was observed to submerge, or when a previously valid radar contact disappeared, the point of submergence was marked and an initial pass over that point made. If contact was not made, a trapping circle was established about the point of submergence. The trapping circle pattern was flown such that the aircraft completes one circuit of the pattern in 3 minutes' time. This means that the size of the circular pattern depends on the speed of the aircraft. Additionally, the length of time to continue to search in this fashion also depends on the speed of the aircraft. An aircraft that searches at 110 knots circles for 47 minutes, while an aircraft that searches at 150 knots circles for 65 minutes.[26]

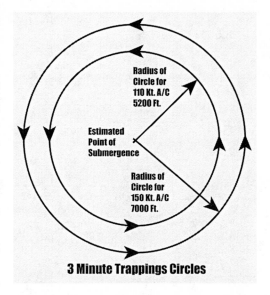

3 Minute Trappings Circles

Diagram of the Trapping Circle pattern used to establish Magnetic Airborne Detector contact on a submerged submarine.

If contact is established by the trapping circle pattern, or the initial pass following submergence, the tracking procedure described below is initiated. The following diagram shows the flight maneuver for transitioning from the trapping circle search to the initial pass of the Cloverleaf Pattern.

Tracking of the submerged submarine was accomplished using the following Cloverleaf Pattern. Using this pattern, the course and speed of the submarine could be determined in order to initiate an attack on the submarine. Alternatively, this procedure might be used to track the submarine while waiting for other forces to arrive. The following diagram illustrates the MAD Cloverleaf tracking operation.

The sequence of the MAD Cloverleaf Pattern is that upon entry the submarine is followed to be detected at Contact 1, which is followed by turn #1 to the left to cross the expected path with detection at Contact 2, which is followed by turn #2 to the left along the reverse to the entry path for detection at Contact 3, followed by turn #3 to the left to again cross the expected path with detection at Contact 4, followed by turn #4 to the left to follow the path of the submarine to a point of launching weapons.

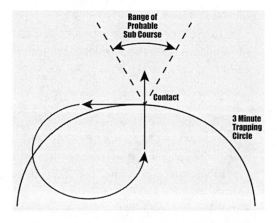

Diagram of the transition from the Tracking Circle pattern to the initial leg of the Cloverleaf Pattern.

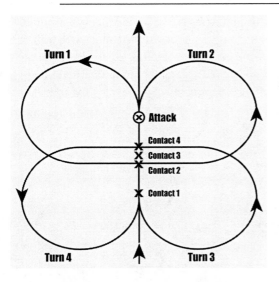

Diagram of the MAD Cloverleaf Pattern.

By being on the same heading as the submarine on the initial entry, after turn #4, and on the reverse course after turn #2 provided opportunity for the aircraft to become better aligned with the course of the submarine, resulting in a better attack.

USE OF ILLUMINATION

When a submarine was located on the surface at night, illumination was used to either identify or attack the submarine. Even if that illumination should expose friendly ships, that risk might be taken if it enabled the aircraft to attack the hostile submarine. The use of this illumination required a high degree of coordination between the various crew members because these procedures are essentially a combination of flight by reference to the aircraft instruments and visual conditions while the illumination is active. Locating and identifying the target and subsequently recovering from the visual conditions back to flight on instruments was very demanding, particularly for the pilot and copilot.

If the illumination method available was the parachute flare, the flight path was selected to move the aircraft from the point of dropping the flare into a position to search the area illuminated by the flare. This helps protect the aircraft and achieve a greater degree of surprise by preventing the flare from illuminating the attacking aircraft. The flare had to be released within approximately ½ mile of the target so that it could provide adequate illumination for identification and the attack that might follow. This required the crew to fly near the target submarine, drop the flare, and then move away briefly while the flare descended.

The flares in use at the end of World War II each had delayed ignition, which allowed the aircraft time to move away from the immediate area of the intense light generated by the flare. It was necessary that the flare be deployed at sufficient altitude so that after descending for the delay time, the flare would still be at a useful altitude from which it would provide the illumination for the crew to search with while it descended further. The following table illustrates the flares available at this time, and their key parameters[27]:

Flare Type	Delay	Used on Aircraft	Release Altitude
Mk 8	90 sec.	PV, TBF	2000 ft.
Mk 8	120 sec.	PBY, PBM, PB4Y	2000 ft
Mk 5 or 6	300 ft.	All	1500 ft

Table of illumination flares available during World War II.

The aircraft would be flown at 1500 ft. for the Mk 5 or 6 and at 2000 ft. for the Mk 8 when searching and flare use was planned. It would be up to the pilot whether to climb to this altitude for flare release for the case when the weather conditions did not provide this much ceiling or cloud clearance.

After obtaining contact with the submarine on the radar, the aircraft is turned immediately towards the point of contact, and homing on the target with the radar was begun. The relative motion of the target, including wind effects, is determined so that a course can be established to pass through a point ¼ mile from the target along the direction of the relative motion. This is illustrated in the following diagram.

This flight path and time were adjusted slightly for a flare of slightly longer or slightly shorter delay time. The pattern was intended to bring the aircraft into a position to search in the illuminated area while the flare was lit, and allow conversion of that flight path into an attack once the target was identified. Most flares provided illumination for approximately 2 minutes, so that the aircraft would remain within approximately 2½ miles of the target during the time the flare descends.

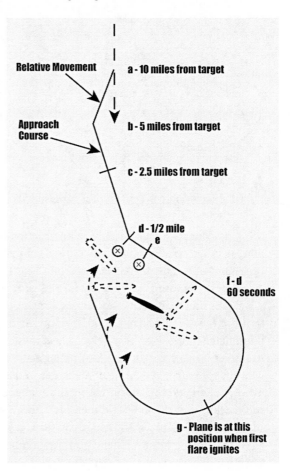

If the illumination method available was the searchlight, the procedure was slightly different. The combination of radar detection and attack by use of the searchlight was the most successful technique for night attack on enemy submarines. The search altitude would not exceed 1500 feet. This provided for optimal use of the detection equipment and allowed the aircraft to reach the approach altitude in a short time. After radar contact was established, the aircraft would descend to 200 feet as quickly as practical while maintaining radar contact and using the radar return to home onto the target. The searchlight was turned on at a distance ½ to 1 mile

Diagram of the pattern flown to observe after dropping flares.

from the target, which allowed for the final heading to be adjusted as the attack was initiated. When the searchlight was turned on, the pilot goes into contact flight procedures, and conducted the attack visually. When the weapons were launched, the light is extinguished and the pilot performed an instrument recovery. There is no need to move the aircraft away from the point of illumination; the source was the aircraft, and so when the searchlight was turned on, the position of the aircraft was identified for any hostile unit in the area. There was also no need to maneuver the aircraft waiting for the illumination. When the pilot was ready for the illumination, it could be turned on.

The following diagram illustrates a pattern of employing the searchlight, after an initial detection by radar. The diagram depicts the aircraft at positions a, b, and c and the contact at positions A, B, and C. After initial detection at point a, the aircraft was turned to point at the detection spot of point A. For half the distance to the target, the aircraft was kept on this heading. At point b the drift was measured so that the heading angle can be increased by twice this measured drift angle until the aircraft reaches point c. The searchlight was trained to the angle of the drift amount up wind of the track and at point c switched on and adjusted to locate the target.

This pattern was intended to bring the aircraft into a position to perform the illuminated search, and allow conversion of that flight path into an attack once the target was identified. Because the source of the light was the searching aircraft itself, the crew was well motivated to use the light for the shortest possible time and limit their exposure to the hostile force.

Diagram of legs flown to employ the airborne searchlight.

ATTACK ON A SUBMARINE

The preceding sections illustrate the various methods available to determine the location of a submarine so that it could be attacked if needed. Now the methods of the attack itself are discussed. In the latter days of World War II, attacks would be initiated immediately on visual identification of an enemy submarine to take full advantage of the surprise factor.

Normally the attack was made on the most direct path to the submarine at the maximum speed of the aircraft. However, if the aircraft is able to descend through clouds, even greater surprise might be possible. To provide greater protection to the aircraft from antiaircraft (AA) fire from the target, during the initial portion of the attack moderate weaving in both direction and altitude could make it more difficult for the AA gunners. Straight flight from a distance of about 1000 yards out provided a stable platform for the weapon release.

Antisubmarine aircraft were equipped with some combination of machine guns, depth bombs, and rockets. Attack with machine guns was directed against the personnel of the submarine or secondarily against the air induction plumbing for the diesel engines, located at the base of the conning tower. A third target for guns were the saddle fuel tanks located abreast the conning tower. During daytime, all guns that were available would open fire as soon as the target was within range. At night, firing the guns was withheld until it was certain that the aircraft had been detected. This would preserve the surprise element to the highest degree possible.

The most common weapon type during this period was the depth bomb. While an attack along the line of the submarine would yield the highest probability of destruction, this course should not have been selected over a direct approach from the point of identification to the submarine. The depth bombs were dropped as a train of several bombs, normally not fewer than four, at the spacing determined from the table below for the case being flown. Each aircraft would carry a minimum of four bombs, or as many more as was consistent with the aircraft in use and the mission limitations. The time interval from the time the depth bomb hits the water until the plume of water peaks following detonation was approximately 15 seconds. The depth bombs would be launched from approximately 100 feet altitude, or lower when practical.

The point of aim for the depth bombs would be such that the center of the stick of bombs to be released would explode at the conning tower of the submarine. If the submarine was on the surface, the bow was used as the point of aim.

The following diagram illustrates these points which were considered in determining where to place the bombs.

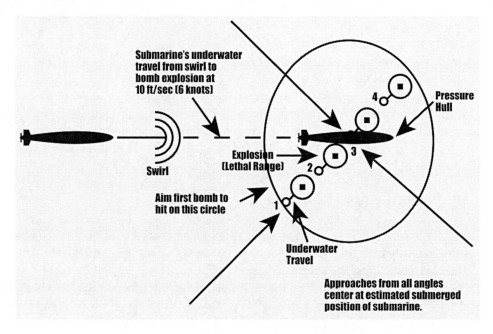

Diagram of an attack on a submarine using depth bombs.

The following table shows the various types of depth bombs available near the end of World War II, applied to the two principal types of enemy submarines of World War II. The resulting diameter of the circle in the above diagram, which was to contain all four bombs shown, would begin either 120, 140, 160 feet from the submarine.

Bomb Type	Charge Weight	Charge Type	Total Weight	Lethal Range (ft)	Impact Spacing (ft)		Attack Circle Range (ft)	
					German	Japanese	German	Japanese
AN-Mk 17	243	T.N.T.	344	19.25	60	70	120	140
AN-Mk41	227	T.N.T.	330	17.25	60	70	120	140
AN-Mk-47	252	Torpex	354	24.0	70	80	140	160
Mark 54	250	Torpex	354	23.8	70	80	140	160

Table of the types of Depth Bombs available during World War II.[28]

The impact spacing of the bombs following release from the aircraft was a function of the angle of glide, altitude, and airspeed of the aircraft at the point of release.

The intervalometer was an instrument in the aircraft that generated a string of pulses a fixed time apart to release the individual bombs. The following table was used to determine the intervalometer setting for the airspeed of the desired release to create an impact spacing of 70 feet. If a 60-foot or 80-foot impact spacing was required, the value from the table was multiplied by 60/70 or 80/70, respectively.

Aircraft Speed in Knots

Angle & Altitude at Release of First Bomb	*100*	*120*	*140*	*160*	*180*	*200*	*220*	*240*	*260*	*280*	*300*
5°—500 Ft.	75	76	77	79	81	83	84	87	89	90	92
5°—300 Ft.	79	80	82	85	88	90	91	93	96	98	101
5°—200 Ft.	81	84	86	89	91	94	97	99	102	107	110
10°—500 Ft.	84	88	90	93	97	101	107	110	116	121	128
10°—300 Ft.	89	94	100	107	111	119	124	131	139	148	154
10°—200 Ft.	96	101	110	118	127	135	145	154	165	176	188
15°—500 Ft.	96	101	108	113	121	130	139	148	158	168	181
15°—300 Ft.	105	112	122	133	145	159	173	190	208	227	248
20°—500 Ft	108	118	128	139	150	165	181	198	218	235	256

Intervalometer Setting for 70-Foot Impact Spacing in Feet.[29]

Some aircraft, both bomber and fighter types, were equipped with forward-firing rockets. These rockets were effective against the submarine in any case where the hull or the conning tower was visible. Although a rocket attack could be very effective, it would not be done in place of an attack by depth bomb. The most effective approach path for the rocket attack was directly abeam the submarine. An approach line that was more than 45° from the beam was not recommended as being much less effective. The rocket attack was made with a glide angle of between 10° and 20° to obtain the necessary horizontal underwater path for the rockets after they hit the surface. The aim point for the rockets was a point 20 feet short of the submarine's waterline beneath the conning tower.

The preceding sections have presented the antisubmarine methods for both aircraft and ship platforms from the time frame of the last days of World War II. This has presented the maturity of both the procedures and the state of art of the weapons and sensors. Generally, these tactics and methods would carry over into the early Cold War period.

TRANSITION TO COLD WAR OPERATIONS

The experience gained in World War II was that of protecting the cargo carrying convoys, and the naval fleets transiting between North America and the far-flung battlefields of the offensive war. To take the attack to the enemy, there was preventative searching for enemy submarines so that they could be attacked before they could carry out an attack. The weapons of the day would allow attack on a ship target from a few hundred yards from that target. Additionally, there were short-term isolations of ports, and landing beaches that required security from enemy submarines. Periods where these areas needed protection was limited to a few weeks time. This significantly limited the size of the defensive perimeter that needed to be maintained around the targets.

In the post–World War II period, although the key technologies and methods of employment had been established by the preceding experience, the need for improvements was recognized. The following areas saw improvements that resulted in incremental innovations to the equipment and use:

1. Widespread use of sound recording of signals to facilitate shore-based replay of missions to backup the work of the search crew.

2. Longer range everything: Ships, airplanes, helicopters, sonars, sonobuoy detections, and weapons.

3. The integrated use of friendly submarines to move the searching into the very water the hostile submarine could occupy.

4. Greater capability and use of communications links to disseminate information to a larger audience sooner.

With the momentum moving from the hot war, when any enemy submarine could be attacked without further characterization, to the Cold War, several important capabilities began to emerge for the antisubmarine forces of the U.S. There was a need to characterize the intentions of an unknown submarine. A submarine being at sea was certainly not in itself an act requiring a specific military response. In fact, both law and custom of the day allowed a submarine to be within 3 miles, and later 12 miles, of the shore with no particular response required. The U.S. response to this new reality was to establish tracking centers for both the Atlantic and Pacific Oceans. Tracking had to be continuous, otherwise a submarine could already be in position when other indications of an attack became evident. In these centers, located in Virginia and Hawaii respectively for the U.S., reports of submarine locations were collected and an up-to-date situation map was maintained. As the ballistic missile-launching submarines came into use by the Soviet Union, the desire to monitor the port areas of that nation to identify the departure of such a submarine from their home waters was begun. The existence of Soviet naval facilities on their Pacific coast and on the Kola Peninsula in the north brought about the need to track submarines entering the Pacific and Atlantic Oceans. These changes led to an increased reliance on the passive sonar methods, so as to not advertise one's presence to the adversary.

Long-Range Detection

Previously, defense against submarine attack was limited to the area immediately surrounding a group of ships. A new need that presented itself during the Cold War was that of detecting the presence of and tracking a submarine not in the vicinity of a fleet needing protection. This need arrived with the inception of the missile-carrying submarine that could launch weapons from the sea. In 1957 it was learned that the Soviet Union had built submarines to do exactly that.[30] The Soviet Union deployed missile-carrying submarines in the mid–1950s, with missiles initially capable of a range of hundreds of miles. This missile range increased over time to 2000 miles. The Soviet submarines had to move well away from their coastal zones and take up positions within this distance, so that they could reach targets within the North American land mass. This made much of the North Atlantic and North Pacific Oceans, areas where these submarines had to be located and tracked, in order to be prepared to attack the hostile boats if ever needed. The long-range detection method of World War II was radio direction finding (RDF) applied to radio transmissions from the submarine. RDF was not available when the submarine made no transmissions.

This range element brought about another subject that grew in importance in the Cold War, that being the range at which the potentially hostile submarines needed to be detected. Monitoring this amount of ocean area was more than could be practically accomplished by ships and patrolling aircraft alone. There needed to be a system of fixed-base sensors, searching outward from the coastlines, and capable of detecting a transiting submarine which could be investigated by the long-range aircraft, or the ships and planes of the anti-submarine fleet.

During and just after World War II, research projects had determined that:

1. Low-frequency sound propagated long distances under water, in most ocean areas. This became known as the deep sound channel.

2. A diesel submarine generated constant frequencies which could potentially be sensed at long distance.

In 1949, Navy experiments with these two findings achieved submarine detections at hundreds of miles from the sensors. In this timeframe, it was suggested that an approach similar to that of the Visual Speech Analyzer, developed at Bell Telephone Laboratories by Dr. Ralph Potter, could be useful in the submarine detection problem.

In 1950, the Navy asked that an industry/academic/government group convene to study and make recommendations on the security of overseas transport, the activity most at risk from the perceived submarine threat. Project Hartwell delivered a report later that year that recommended the use of low-frequency spectral analysis for submarine surveillance. Other work in both university and government lab settings provided additional evidence of the usefulness of these concepts. The Navy contracted with Bell Laboratories for Project Jezebel to demonstrate these concepts. In 1951 Bell delivered a brass-board model of the Low Frequency Analysis and Recording (LOFAR) system. This was installed in a test site on Eleuthera in the Bahamas. The installation consisted of arrays of hydrophones located near the coast, connected to a shore-based facility for processing. In 1952 demonstrations of this test installation convinced Navy officials of the value of the developments.[31] This led to initiating Project Caesar to establish six installations to field the equipment manufactured based on these contracts. The sites chosen were in the Atlantic Ocean adjacent to the U.S. coast: Bermuda, Eleuthera, Cape Hatteras, and Puerto Rico. In 1952 this system of sensors and receiving stations was termed SOSUS for SOund SUrveillance System. A SOSUS sensor consisted of an array of passive hydrophones, located on the sea bed, and attached by cable to a shore-based processing facility. Detections by these facilities could be made of a submarine as much as 3500 miles from the sensor arrays. The primary use scenario for these stations is to provide an indication and identify a relatively small area to be searched by the mobile forces described above. Often a report from one of these shore-based stations would lead to an assignment of a long-range patrol aircraft to conduct a search of the identified area, using any or all of the sensor systems available on that aircraft. There were a number of these sets of connected hydrophones covering both the North Atlantic and North Pacific Oceans. These initial sites were among those that based the sensing equipment for thirty years or more. These stations, called NAVFACs, for naval facilities, began coming on line in 1954. As the network evolved, there came to be twenty-three such stations covering the northern areas of both the Atlantic and Pacific Oceans. When needed, the cover story for these closely held secret sites was that they were oceanic research stations. The Navy only acknowledged these sites and their importance in 1990.

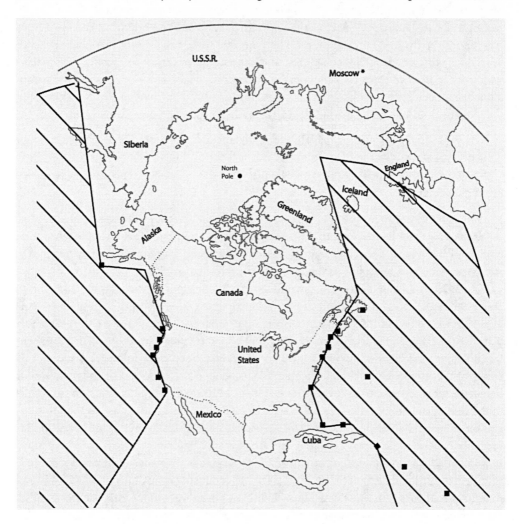

Map of SOSUS stations and the detection coverage they provide.

A SOSUS installation of multiple hydrophones uses the technique of interferometry to combine the signals to produce an estimate of the direction of arrival. These SOSUS networks required the invention of a new technique to extract an extremely small signal, such as from a distant submarine, amid a background of other ocean noises. This process was LOFAR, for Low Frequency Analysis and Recording. Listening to the sounds picked up on hydrophones in general, and specifically the sensing element of the sonobuoys, allowed for no accumulation of signal content. You could only listen to a particular sound at the moment of reception; if the small desired signal was completely overwhelmed by the noise content, the signal could not be heard. With LOFAR, the time-based sound signal was converted to a frequency-based signal, or the set of frequencies making up the sound.

Once converted to this frequency-orientation, brings in the concept of a frequency bin. This concept is very important to extracting signal information from a noisy background. The significance is that the frequency components of a signal do not move around. The second major concept involved here is the random nature of the surrounding noise of the signal heard on the hydrophone. One of the characteristics of random noise is that it sums to zero over time. By adding within these frequency bins, the signals that are persistent, because there is a signal present at that frequency, and the random noise elements from non-persistent sources, the sum of these signals and noise will be the signals.[32] The noisier the signal, the longer the summation of signal and noise is required to remove the noise.

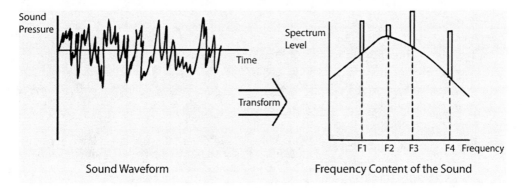

Depiction of LOFAR process.

The recording element of the acronym was implemented by the printing of this frequency-based signal, or set of frequencies represented by the frequency bins generated by a particular set of equipment. This resulted in printing of the intensity of LOFAR frequency bins, often called the LOFAR gram, by a printer device usually known as the "gram printer." This printer was somewhat like an early computer printer, in which a print head moves across the paper and at each character position makes the symbol for the character to be printed; much like the individual characters on the lines of this page. The LOFAR gram printer instead used the line of paper as a line of individual dots, each dot corresponding to a particular frequency bin of the processed hydrophone signal. For each dot on the line, the printer would make a lighter or darker dot corresponding to lesser or greater intensity of the group of frequencies of that bin. The line of dots would begin on the left with the lowest frequency detectable by the equipment and progress to the right edge with the highest frequency detectable by the equipment. Again, those frequencies that did not persist would show up as a darker dot occasionally and would largely fill the page with a random collection of dots representing the background noise of the signal. However, if a persistent frequency was present, such as from a submarine or ship's propeller blade, or from an engine on a ship or submarine, these frequencies would persist and would show up on the page as a line of darker dots, indicating that a frequency in this bin was persisting. The operators who monitored these printers were taught that natural sounds, including those from animals, contained many frequencies, and as a result would not show up on the printed sheets as individual lines. However, man-made sounds such as those from machines would show up as individual frequencies. The printer generated a line of dots, say every second, and the sheet of paper already printed represented numerous seconds of such printing. This allowed the human operator to judge the degree of presence of a particular frequency, even though the actual received signal was completely obscured by the noise if he listened to the audio sound from that sensor. This LOFAR processing of the sounds received from the hydrophones was the key to both aircraft use and shore-based usage to greatly extend the range from the receiving hydrophone to a distant source of sound. One difference between the airborne equipment and the shore-based facility was that following introduction in the aircraft, each aircraft was able to monitor up to four such hydrophones simultaneously, whereas the shore-based SOSUS processing building was able to monitor perhaps a hundred such signals simultaneously.

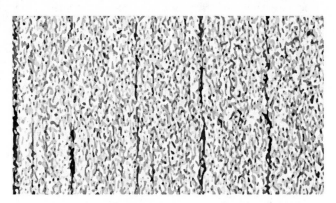

An example LOFAR "gram" showing darker lines for persistent frequencies, and lighter marks for random and non-persistent background noise.

ASW PROCESS

The previous wartime need to counter enemy submarines in the vicinity of a fleet of ships, around ports, and around areas such as landing beaches was adapted and expanded to monitor non-friendly submarine activity in both the Atlantic and Pacific Oceans in the Cold War. This monitoring provided awareness to national authorities of the potential threat of those non-friendly submarines at any given time. This meant establishing the means of using the long-range sensors to detect the existence of a submarine, along with response forces to investigate essentially any report or detection. These means became steps in the ASW process.[33]

The ASW process involves moving through a series of actions: Detection, Localization, Classification, and Prosecution. Depending on the circumstances, progression from one action as prelude to the next is skipped or simply not done. For example, if an aircraft were sent to relieve another aircraft which was running low on fuel, the detection, localization, and classification might not be needed as the previous aircraft can direct the arriving aircraft precisely to the proper location. And in peacetime, a prosecution employing weapons might simply be inappropriate.

Detection and Localization

The first two steps in the ASW process are detection and localization. Detection is the accomplishment of finding something of interest in a general area that is worthy of further examination. Localization is the confinement of the object's position to a workable area. Detection is normally by the use of one of the long-range sensors: SOSUS, radar, ESM, or non-directional acoustic sonobuoys. With the knowledge that something of interest is "out there," additional sensor resources are applied to determine more precisely where and what the object is. This step would often entail using sonobuoys to determine direction to the object, and/or one of the imaging sensors, and/or the magnetic anomaly (MAD) sensor.

Classification

The next step in the ASW process is the classification of a detected object. Determining whether submarine or non-submarine is the key differentiation between objects. If this cannot be accomplished visually, other uses of the available sensors are made. Classification of a submarine is done in one of three ways: observing a set of specific frequencies associated with a particular submarine; a visual sighting; or the presence of an underwater metal object with nothing visible at the surface. Note: Although this step is presented following localization, the sequence is less important. The earliest possible classification of the object is the objective, and may in fact render further localization unneeded.

Tracking

The next step in the ASW process is tracking the object, if needed. Tracking is learning the direction and speed of the submarine, so that a prediction of its location at a later time can be made. This may be thought of on two levels; following localization, tracking may be required prior to weapons employment. Alternatively, tracking on a larger scale may be the objective of the detection and localization. The objective may be to locate a vessel, or other object, so as to keep track of its location over time; for example, a submarine or ship may be relocated after one or more days' time have elapsed since the previous location in order to predict its final destination. Such tracking goes a long way toward determining the intent of the submarine or ship.

Prosecution

This step involves employing weapons.

Naval Fleet to Find Submarines

The term Hunter-Killer began to be used in World War II to label a unit that was capable of both hunting and killing the prey, if necessary and appropriate. The ability of a single unit, even an individual aircraft or ship, to do both tasks replaced the older methods of having separate patrol or reconnaissance functions and attack units. These reconnaissance functions reported what they found, including hostiles perhaps in need of destruction, and the attack function, equipped to destroy particular classes of target once identified, responded. The problem created by separating the two tasks into separate units, with separate and different equipment, was that the information had to be retrieved from the reconnaissance element and transferred to an attack element launched with the information. If the target was a rail line, bridge, marshaling yard, bivouac area, dam, mill, or virtually any other object relatively fixed in location, this approach worked well. Once the target became the highly maneuverable submarine, which could hide completely at least for a period of time, this approach had serious limitations. The solution to this was the hunter-killer unit. The key element the hunter-killer units possessed was the ability to move from the search phase of hunting to the attack phase in a very short time, perhaps even seconds in some circumstances.

The first generation of submarine-hunting defense after the war was the Hunter Killer Group.[34] These Hunter Killer Groups were the front line antisubmarine response force from about 1954 to 1963. The Hunter Killer Group was comprised of an aircraft carrier, initially of the Essex class, which provided a mobile base for both fixed-wing and helicopter aircraft, and several destroyers. Of the destroyers, some of these ships were equipped to provide defense of the fleet against air attack, while some were equipped to provide defense against attack by submarine. During this timeframe a single destroyer ship did not perform both functions, as there was only room on a single ship for the weapons of one of the functions. These groups could be dispatched to any area of ocean where a potentially hostile submarine was thought to be or reported to be operating. Three such groups shared the duties for the Pacific Ocean and three more for the Atlantic Ocean. In each ocean one of these groups was at sea at any time during peacetime, and perhaps more than one as tensions increased.

The destroyer ships of the Buckley class (DE) provided active sonar, both air and surface search radar, weapons such as homing torpedoes, antisubmarine mortars known by the name Hedgehog, and traditional naval armament such as the 5-inch deck gun and depth charges. Destroyers could be dispatched from the central carrier to investigate a report some distance from the group. These ships moved at 15–25 knots.

The fixed-wing aircraft of the Hunter Killer Group consisted of three types: initially the Douglas AD-5, to be replaced by the Grumman E-1 Tracer radar aircraft, and the Grumman S-2 Tracker antisubmarine aircraft. The helicopters of this era were the Sikorsky HSS-1 Sea Bat, which was relabeled the SH-34 in 1962.

The Grumman E-1B Tracer radar plane operates far ahead of the task force, using its radar to continually search the ocean surface for reflections from a hidden submarine.

The Grumman Tracker is the primary antisubmarine long-range detection and localization airborne system assigned to the Hunter Killer Group. The S-2 Tracker antisubmarine aircraft was powered by two radial engines and carried a crew of four, made up of pilot, copilot, radar and MAD operator, and the acoustic system operator. The operational scenario of the S-2 was to remain on station in a target area for five hours at a distance of 250 nautical miles from its carrier base. Its cruising speed was 150 mph with a range of 1350 miles or 9 hours' endurance. This aircraft contained the systems for search and detection, as well as for killing. It had an internal bomb bay that could carry a combination of weapons including homing torpedoes, depth charges, and a nuclear depth bomb. Antisubmarine sensor equipment included up to 32 sonobuoys launched from the rear section of the engine nacelles, AQA-3 or -4 Jezebel passive sonobuoy detection sets, ASA-20 Julie plotter/recorders, a Magnetic Anomaly Detector (MAD), an electronic support measures (ESM) detection set, and the APS-38 search radar. Supporting the sensor operations were 60 explosive charges carried on the airplane, to create a sound pulse heard on the passive listening equipment, when reflected by a nearby object in the water.

The Sikorsky HSS-1 Sea Bat antisubmarine helicopter also carried a crew of four: pilot, copilot, radar and MAD operator, and the acoustic operator. This radial engine–powered helicopter had a range of 180 miles and a speed of 123 mph. It carried the AQS-4 (or -5) dipping sonar and homing torpedoes and depth charges for weapons. Once in the contact area, the helicopter could be put into an automatic hovering mode, 40 feet from the water. From this perch it could lower its sonar transducer, the dipping sonar, to search for what lies beneath. Sonar operators onboard the helicopters listen to and watch their instruments for indications of anything below the surface. These helicopters, like the S-2, could both locate the elusive target and attack and destroy that target.

The long-range land-based patrol aircraft of the era was the Lockheed P-2. This purpose-built twin radial engine airplane normally carried a crew of 10, including pilot, copilot, a Jezebel operator, a navigator, a tactical coordinator, a Julie/ECM operator, a radar technician, a radio operator, a MAD-ASR operator, and an ordnance technician. The radio operator and the ordnance man also served as observers stationed in the rear of the aircraft, and the MAD-ASR operator served as observer in the nose of the aircraft. This aircraft had a cruise speed of 170 knots and a patrol endurance of 4.5 hours at a distance of 500 nautical miles from its shore-based runway, or endurance of 8 hours at 250 miles distant. Said another way, with full fuel the aircraft had a maximum of 12 hours' endurance, and you could use this endurance more than one way. In addition to the sensor systems similar to those of the S-2 above, the P-2 was equipped with a diesel sniffer to detect the fumes of a diesel motor.

Use of Dipping Sonar

The sensor systems of the aircraft were nearly identical in concept to those of late World War II, although the helicopter-borne dipping sonar was new. The sensor ball was attached to the helicopter by a cable and was lowered down into the water and then hoisted up for normal flight. A unique capability of the dipping sonar ball, when in the water below the hovering helicopter, is that it is free from surrounding man-made noise, effectively increasing its range to a submerged submarine. This sensor is an active sonar that sends out a sonar pulse, to be returned by an object in the water. Its secondary use is to listen only to pick up sounds in the water.

The dipping sonar is most often used by two helicopters in concert to search an area. While one helicopter has its sonar ball in the water, the other helicopter is moving forward to the next point for the sensing. As a result of this leapfrog operation, one helicopter always has the sensor ball in the water as the two helicopters move across the area.[35]

ECM

Electronic Countermeasures (ECM) sensing instruments were added to the P-2 and S-2 aircraft. ECM as a discipline includes the sensing of signals used by an adversary force, as well as actions taken to deny the effective use of those signals by the adversary. The P-2 aircraft did not have any means of countering, that is interfering with, the transmitters it could sense, but the equipment and its operation were labeled as ECM because the equipment was common to other ECM platforms. The equipment responded to electronic transmissions most commonly associated with radars. Location of search radars and tracking radars associated with defense systems were the primary objective of this equipment.

The antisubmarine application of this equipment was to establish the direction from which the radar of a submarine was transmitting. The line-of-bearing obtained would provide the crew a direction to search. For example, if a submarine exposed an antenna above the ocean surface in order to operate its air search radar, likely the searching aircraft's radar could detect it as well. Additionally, the ECM sensing was passive; it produced no signal the adversary could use to locate the searching aircraft.

This equipment aboard the aircraft also established a secondary mission for the plane and crew, that of identifying and cataloging the locations of radar transmitters in general. In addition to those located on submarines, those located on other naval vessels and those located on shore could be located. For example, the aircraft could be flown along the coast-line of a country, remaining in international waters, and identify any radar transmissions sensed along that coastline. Once a radar transmitter was identified, its signal would be reexamined periodically, say every five minutes, to record a number of lines-of-bearing to provide the means of locating the transmitter through the cross bearings.[36]

The set of ECM equipment on the P-2 consisted of two ALR-3 countermeasures receivers, an ALR-8 countermeasures receiver, the APR-69 azimuth panoramic display unit, and the APR-74 pulse analyzer. Each ALR-3 receiver is scanning over a frequency range of 2,300 to 4,450 or from 7,050 to 10,750 MHz, the frequency bands used for radar at the time. These ALR-3 receivers are used to identify the existence of a signal and its general location in the frequency spectrum. With an indication from this receiver, the operator tunes the ALR-8 receiver to this area of frequencies for closer examination. After adjustment of the ALR-8 receiver, this signal can be viewed on the APR-69 and APR-74 displays. The APR-69 performs automatic direction finding to the transmitter or indicates the frequency in panoramic display. From this direction finding the operator extracts a bearing to the transmitter. The APR-74 signal analyzer shows the signal on a CRT display as five lines, each with a separate time span. With these multiple lines displaying the same signal, the operator extracts the specific characteristic parameters for the signal. The pulse width and pulse repetition times are extracted from this display. By using a stopwatch, the operator would determine the modulation pattern of the transmitter to complete the set of infor-mation to be recorded. For example, if a transmitter were observed to be present for a very short time every 12 seconds, the tentative conclusion would be that it is an air search radar rotating at 5 RPM. These parameters are recorded on paper. Over time, a file of known sig-nals and their locations, or source, is built up so that on a given flight, the ECM operator may be verifying previously known signals but is also on the alert for unknown transmis-sions.

Increased Use of Sonobuoys

In the Cold War, the use of sonobuoys to determine the location of a submarine was increased due to the need to detect the submarine at a time and place where it would not reveal itself, as had been the case in World War II. The increased use of the snorkel by sub-marines meant the submarine was beneath the surface more of the time, requiring increased sensing beneath.

The early sonobuoys were omnidirectional and provided only the sound received at the buoy location. The following pattern of buoys was established to localize a suspected submarine. The aircraft released the purple buoy at the point of the suspected submarine, then two miles later it releases the orange buoy. After flying a turn to the left, it crosses the purple buoy, then two miles later it releases the blue buoy. After flying another turn to the left it, crosses the purple buoy, then two miles later it releases the red buoy. After flying a final turn to the left, it crosses the purple buoy, then two miles later it releases the yellow buoy. This entire procedure takes about thirteen minutes to complete.

The range of these early sonobuoys varied from as high as three and a half miles down to very small distances.

The methods of utilizing the sonobuoys to detect or identify and then localize a submerged submarine were known as Comparative Listening, Jezebel, Julie, and CODAR. Jezebel and Julie were both named for women of less than angelic character. In the Bible, Jezebel is the wife of Ahab, generally known as a schemer and for wickedness. Julie is a more modern reference to burlesque dancer Julie Gibson, a

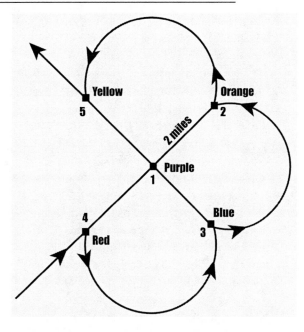

Diagram of the POBRY sonobuoy pattern of the early Cold War.

well-known entertainer in a 1950s Philadelphia nightclub whose "Dance of the Bashful Bride" was said to turn passive boys (buoys) active. CODAR is the acronym for COrrelation Direction and Ranging.

COMPARATIVE LISTENING

The trained operator would gauge the level of sound produced by each sonobuoy in the water and judge the general location and source of the sounds. Changes in the frequency of the sounds could be interpreted by the operator as up or down Doppler shifts in the frequency, indicating that the source was either moving toward or away from the receiving sonobuoy.

At the end of World War II, and into the 1950s, antisubmarine aircraft would use sonobuoys, normally in a pattern of five buoys, to passively determine the location and movement of an adversary submarine. These were placed at equal distances apart to facilitate the comparative listening technique. The operator would listen in turn to each of the buoys and listen for the difference in sounds heard on each buoy. In addition to the differences in intensity of the sound on each buoy, they were listening for specific components of sound, such as the moving propeller, movement of bow planes, other machinery such as motors, and even noises caused by the crew. The operators had learned from their training that each of these types of noise could be heard when the source of the sound was within a certain distance from the listening sonobuoy. Comparative listening leads to using the difference in the intensity and quality of the sound on each buoy to predict the range from that buoy. By assigning a number from 1 to 10, with 1 being a very strong signal, for the sound from each buoy, a relative position could be determined with this technique;

when applied to the pattern of buoys, it would allow picturing the source of that sound within the pattern of buoys.

A relative position is marked where the sound from the purple buoy at level 7, the sound from the blue buoy is at level 6, and the sound from the red buoy is at level 8. Using this technique to deduce a position at a different time would allow estimation of the motion of the sound source over that time.

With the equipment of the time, this comparative listening was primarily limited to a range from each sonobuoy of about 2000 yards. Beyond this range sounds, such as explosions could be heard, but not sufficient detail of the sound to distinguish those sounds specific to a submarine.

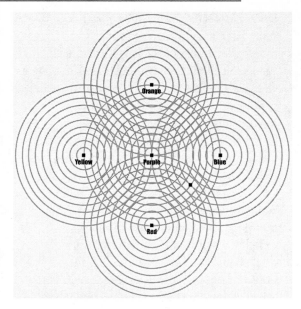

Diagram of concentric circles about the pattern of sonobuoys used for estimating the location of a sound.

JEZEBEL

The change in the use of the sonobuoys in this period was to decrease the reliance on an operator listening to a single buoy, and replace this approach with the monitoring of the set of buoys through equipment designed to aid in that task. New equipment was added on the aircraft to augment listening to sonobuoys. Four sonobuoys could be monitored simultaneously on the AQA-3 or -4 recorder. These instruments were the beginning of processing of acoustic signals for extraction of the useful information in the submarine location problem. This change also reduced the reliance on the sound recognition skill and experience of the operator.

In Jezebel, the passive sounds received from a sonobuoy, and hence from the surrounding water, were processed for display in a continuously moving paper graph of frequency versus time. Each type of ship would have a consistent set of frequency lines, or tonals, which could be used to identify the source of the machine creating the set of lines received when near that type of ship. The Jezebel operation is termed narrow-band processing, also known as LOFAR, in that individual frequencies are the telltale indication of the target machine. The operator would watch the visible plot of four buoys to see what frequencies were present on the plot of multiple buoys, particularly adjacent buoys, and based on results, recommend additional buoy placement. Changes in those frequencies could be interpreted by the operator to have up or down Doppler shifts in frequency, indicating that the source was either moving toward or away from the receiving sonobuoy.

JULIE

Julie is a process of listening for the return from a submarine, or other object in the water, of a sound pulse initiated for that purpose. Specifically Julie involved first dropping the sonobuoys to listen, and then dropping a small explosive device to create the sound pulse to be reflected off any suitable object in the area. These small explosive devices were called Sound Underwater Signals, or SUS's, and were also known as Practice Depth Charges, or PDCs. This combination of causing the sound of an explosion and monitoring the return signal was later called Explosive Echo Ranging, or EER. This echo-ranging operation is termed broadband processing in that the returning sounds need only be appropriate in frequency range to be received by the equipment in use. The information extracted from this process was the difference in time for the object to reflect the single explosive signal to the two sonobuoys in the water at different locations. These two time delays represented the distance from the object to the receiving sonobuoys. These two times were used to plot the position of the object in the water.

For localization, two sonobuoys would be monitored on an ASA-20 chart recorder that showed the sequence of events in the sound signals. By measuring the distance between the explosion and the reflection by this sound from the target received at the two sonobuoy locations on the trace, and applying the bathythermograph information to predict the speed of sound in the water, the resulting time-delay information could produce an estimate of the position of the object reflecting the sound.

CODAR (LOCALIZATION)

A further use of the passive omnidirectional sonobuoys was CODAR, or COrrellation Direction And Ranging. This was a process whereby the signals from two non-directional sonobuoys could be combined to determine the angle from which the sound pulse was received.

The objective of this pattern is to determine the location of the submarine to a sufficient degree to be able to employ weapons, if that is desired. Two buoys are placed 350 feet apart. Two additional buoys are dropped on a line perpendicular to the line of the first pair.

Electronics on board the airplane processes the signals to determine the angular relationship to a target. Each pair of buoys establishes a baseline. By comparing the phase of the signals received at each buoy from a common point, i.e., a target, a bearing line to the target is determined. The relationship establishing this bearing line is true from either side of the baseline, creating an ambiguity. The ambiguity is

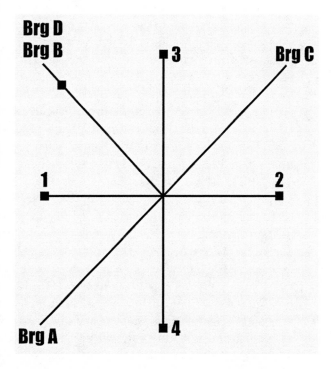

Diagram of the pattern for the four sonobuoys used in a CODAR localization.

resolved by the second pair of buoys and the second resulting baseline. Only one bearing to a target will satisfy the relationships for both baselines, and hence establish one unambiguous line-of-bearing to the target.[37]

An additional problem that had to be faced in the Cold War was operating the sonobuoys in the different oceans. During World War II, it was primarily the North Atlantic Ocean where the sonobuoys were used. With the Cold War, the North Atlantic and the North Pacific required the equipment to be used necessitating understanding the travel of sound in all these different waters. This led to a distinct increase in the amount of ocean research that was being conducted to improve the understanding and develop differences in operating procedures for the crews operating in these varied conditions. The effect of water temperature, salinity, and pressure were studied and modeled so that operating crews could have predictions of the effect of these parameters on the speed of sound in the water where they were working. The simplified effect of these efforts was to enable the crew to determine at what spacing the specific sonobuoys used needed to be placed and at what depth in order to maximize detection of a submarine.

All of the platforms have a quality of reach. For each maneuverable platform—ship, aircraft carrier, helicopter, or airplane—there is a quality of the platform that establishes how quickly it can move from its base to an area of interest, presumably to provide its unique capabilities of sensors and/or weapons to a problem. Second is the time that it can provide these capabilities while in that area. This quality applies equally to a ground-based airfield or shore base of ships from which the mobile elements can be launched to a region of interest.

This concept of reach also applies to the individual sensors and weapons which the mobile platform brings to the area. The significance of the carrier battle group is that the carrier itself can move into a region from which it can launch its aircraft and helicopters to provide a shorter transit time to the specific area for operations. Likewise, the destroyers of the carrier group can detach and move out from the group to a designated area, and in the most modern incarnation of this concept, itself launches a helicopter which can operate out to a certain range from its destroyer base to utilize its sensors, which have a range of effectiveness as well.

The command center of this Hunter Killer Group was housed in the aircraft carrier from which the antisubmarine helicopters and aircraft were launched and recovered. The core of the task force was the aircraft carrier, the quickly moving airbase for the fixed-wing and helicopter aircraft described above. The carrier was the command and communications center of the entire operation that controlled up to four squadrons of aircraft, a squadron of destroyers, and the 5000 men it took to launch, maintain, and operate the aircraft and the sensors they carried. All of these men and equipment were focused on the set of tasks to detect, classify, and destroy, when needed, hostile submarines across thousands of square miles of open water. The central authority of the carrier is the combat information center, or CIC, where all information is evaluated immediately. The decisions made here are dependent upon having a complete and up-to-date picture of the situation. The carrier's radar could be used to direct aircraft, either the land-based P-2 Neptune, the carrier-based S-2 Tracker, the carrier-based SH-34 helicopter, or a combination of them, in a search operation based on a reported contact position.

This combined task force would be assigned to investigate a report of submarine sighting, or it could identify a potential threat through its own long-range sensor, the AD-5 or E-1 Tracer radar plane. This aircraft would typically be stationed some distance ahead of the moving task force, within a nominal range of 250 to 300 miles, sweeping the area for any radar reflection. Its onboard radar, the APS-82, could identify the reflection of a submarine's periscope, or the submarine itself if exposed above the surface. This radar had an airborne moving target indicator feature, which allowed it to distinguish a moving target from the background clutter produced by the ocean's surface. The operational scenario of the E-1 was to orbit for slightly less than five hours at a distance of 150 nautical miles from its carrier base.

The initial report of a submarine could be from several sources. A long-range patrol aircraft, such as the P-2, could detect something with its search radar, the carrier's radar search aircraft could similarly make the detection, the long-range passive sonar system SOSUS could provide the information, or a passing commercial ship or overhead airliner could report a suspected submarine. The report from any of these sources would be routed to the headquarters of the antisubmarine forces for either the Atlantic or Pacific Oceans located at Norfolk, Virginia, or Pearl Harbor, Hawaii, respectively. The reported location of a new contact would be added to the plot of the tracking facility and the nearest ASW task force would likely be directed to the area to investigate.[38]

Once alerted, the activity aboard the aircraft carrier would change to that of preparing either fixed-wing or helicopters, or both, for launch. Aircraft are moved from the below-deck maintenance spaces to the crowded space of the flight deck. Flight crews would receive briefings on the reported threat, the cloud cover, visibility, air and water temperatures, and the search procedures to be used. The helicopters and fixed-wing aircraft are then launched to the contact area.

Following these briefings and other preparations, first to depart would be the SH-34 helicopters, which would move toward the target area at 120 knots. These would be followed by departure of the S-2 aircraft, which would move at 150 knots. Somewhat prior to the departure of the carrier-borne air assets, destroyers sent to investigate would be moving at something less than 25 knots toward the contact area. If the initial contact had been made by a P-2 aircraft, it would remain on the scene monitoring its sonobuoys and other sensors until its fuel level forced it to depart. This could be either before or after the arrival of assistance from the carrier depending on the fuel state of the P-2 at time of initial contact. First to arrive of the approaching forces were the S-2 aircraft. These aircraft would most likely lay a pattern of sonobuoys with the intention of reestablishing contact with the suspected submarine.

Very similar to the concept of reach discussed above for the investigating platforms, the moving target, such as a submarine, can move a distance from the last sighting before the yet-to-arrive search forces can begin looking for it. The last known point of contact for the target includes a time in addition to the position of the contact. The length of time until the search platforms can be in the area applying their sensors provides some upper limit on the distance a target could move. This distance applied to this last reported location, or datum, determines the size of the area the search forces would need to examine when they get into the area. Of course the size of this circular area, since one presumably doesn't know the direction the target may have moved, continues to grow as the search commences and continues. The largest circular region to be searched is established by assuming the fastest possible submarine is involved.

Once the forces arrive in the region of the contact, they begin to deploy their individual sensor systems. Each of these sensor types has its own limitations on use. These search systems are each bounded by the medium in which they can detect. For example, the radio frequencies of the search radar propagate in the air above the ocean, but their propagation beneath that surface is essentially nil. This is also true for the visible and optically aided vision sensor, or human eyeball. This sensor can see objects above the surface but to a very little degree beneath. The sonobuoy is the device unique to submarine hunting that transmits over the VHF radio transmitter above the ocean surface those sounds it detects beneath the ocean surface. This sound energy beneath the surface has essentially no propagation in the air above the surface. The magnetic anomaly detector (MAD) is only sensitive to several hundred feet. Therefore to have maximum detection depth beneath the surface, the sensing aircraft must employ this sensor as close to the surface as practical, usually a few hundred feet altitude at the most, to be effective. As another example, sonobuoys transmit their signal to the monitoring aircraft on the line-of-sight restricted VHF set of radio frequencies, usually with a 1-watt transmitter, sometimes less. So an aircraft operating close to the surface to utilize the MAD sensor would have very limited reception range to sonobuoys around the area, limited by the small line-of-sight VHF reception range while at the low altitude. Conversely, an aircraft that had established a higher altitude so as to monitor several sonobuoy transmitters in the area would not be able to use its MAD sensor at the several thousand feet altitude needed to receive sonobuoys spread over several miles of ocean. Similarly, radar used to detect a submarine or periscope or snorkel tube above the surface would have an optimal altitude for operation where returns from the metallic target would be optimized versus returned signals from the surrounding ocean. The visible or optically aided vision sense is affected by employment altitude. The primary limitation to this vision in a given set of conditions is interference such as clouds, smoke, haze, or precipitation along the slant range, that range from the observer to the object. The radar sensor is also an application of this slant range effect. Near the surface, a much higher percentage of total slant range is the component away from the sensor platform, whereas at higher altitude a portion of the slant range is consumed reaching from the platform to the surface. Whatever slant range from the aircraft is possible by the individual sensor system, the operator wants to take maximum advantage to place the sensor platform as far as possible from whatever danger may exist near the target's location. Despite these limitations of individual sensor systems, the aircraft platform is often the most desired method of moving the sensor into the area, since it can arrive quickly, depart quickly, and overall remain highly mobile and flexible during an investigation.

These searches may go on for hours. Each aircraft type remains in the area for as long as its fuel supply will allow. The destroyer ships are able to stay essentially as long as needed or until directed elsewhere. These methods constituted "hold down" tactics, where the diesel-powered submarine was not allowed sufficient time on the surface to recharge its batteries. With its batteries depleted, the submarine was significantly less of a threat and quite vulnerable.

The year 1957 brought to the equipment and methods of antisubmarine warfare what is considered the best vehicle for detecting and tracking a submarine: the nuclear attack submarine. The submarine, particularly the nuclear-powered type, has tremendous endurance. Its other primary attribute is the high degree of covertness. These two elements combined with its primary sensor, the passive sonar array, allow the submarine to work some distance from the protected force and detect other submarines a considerable distance from the friendly ships, either transport or military.

After this addition to the antisubmarine forces, the sensing and attack platforms available to the ASW group commander consisted of the submarine, the destroyers, and the carrier itself, which also is home to fixed-wing and ASW helicopter aircraft. Rounding out this lineup is the land-based long-range maritime patrol aircraft. All of these platforms have submarine sensing equipment as well as weapons, all of which have evolved and improved through the years.[39] In addition to participation in and providing protection to a carrier-centric battle group, other units of the nuclear-powered attack submarines were assigned to seek out, track, and attack when required the missile-carrying submarines of the Soviet Union. The same qualities of stealth and endurance they brought the carrier battle group were equally applicable to the searching for the strategic submarines of this major presumed adversary.

In this same time frame, Soviet submarine and missile development provided them with the ability to launch nuclear missiles from submarines. This created an immediate threat for U.S. forces to defend against.

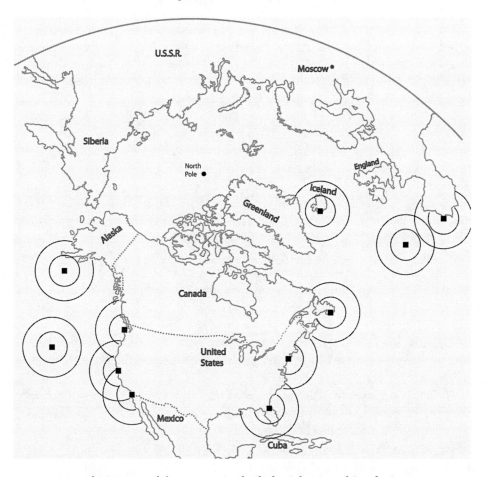

Map of ASW aircraft bases covering both the Atlantic and Pacific Oceans.

The response of the U.S. to the need for response forces for both the Atlantic and Pacific was to station long-range maritime patrol aircraft at air bases along the periphery of both oceans. These bases would enable the investigation of any reported contact within much of both the North Atlantic and the North Pacific Oceans. During the shank of the Cold War period, first the P-2 Neptune, and later the P-3 Orion aircraft were based in this manner. As a result, the in-depth protection against hostile submarines for the continental U.S. was set by having permanently placed hydrophone arrays along the coasts to provide the early warning across the vast open ocean, and mobile forces ready for dispatch. Two types of dispatch forces were at the ready: the long-range patrol aircraft, and the Hunter Killer task force of aircraft carrier and supporting ships, as described previously, carrying both fixed-wing aircraft and helicopters. Any of these responding units were capable of finding and prosecuting an attack on a hostile submarine if and when conditions warranted.

The above map shows the location of these shore-based ASW bases. The two rings about each base location show the nominal operating range of the Lockheed P-2 Neptune as the smaller and the Lockheed P-3 Orion with the larger circles.[40]

The Antisubmarine Task Group

The second generation of submarine-hunting defense after the war was realized in the antisubmarine task group, built around the Forrestal class of aircraft carrier. Although not dramatically different from the Hunter Killer group, evolution in the equipment causes me to draw distinction in the overall capabilities. There was an increased need to detect the submarine at longer range due to Soviet weapon development.

The destroyer ships of the Forrest Sherman class (DD) provided active sonar, and weapons such as homing torpedoes, and traditional naval armament such as the 5-inch deck gun and depth charges. Early in this period these ships were still equipped with the antisubmarine mortars known by the name Hedgehog, but these were removed over time in favor of rocket-propelled weapons. Similarly, the previously used method of launching depth charges to the side and rear of the destroyer was replaced by the newer Weapon Alpha, which could project depth charges 400 to 800 yards out from the ship in a desired pattern. The newer SQS-23 sonar, with detection range out to 10,000 yards, made it necessary for the destroyer to be able to attack a submarine at this range. These Forest Sherman class destroyers were capable of a maximum speed of 32 knots, and had a maximum range of 4000 nm at 20 knots. These ships provided for an air controller to direct the airborne assets of the carrier in a coordinated employment in the area of a suspected hostile submarine.

Replacing the Forrest Sherman destroyers over the next decades were the Farragut, the Knox, the Spruance, the Oliver Hazzard Perry, the Kidd, and the Arleigh Burke classes of destroyers.

The fixed-wing aircraft of this group consisted of two types: the Grumman E-1 Tracer radar aircraft, and the Grumman S-2 Tracker antisubmarine aircraft. The helicopters of this era were the Sikorsky SH-3 Sea King.

The long-range land-based patrol aircraft of the era was the Lockheed P-3. This airplane had four turboprop engines and normally carried a crew of 12 including pilot, copilot, flight engineer, a Jezebel acoustic operator, a tactical coordinator, a navigator, a Julie/ECM operator, radar/MAD operator, a radio operator, an extra pilot, an extra technician and an ordnance technician. It had an internal bomb bay that could carry a combination of weapons including mines, homing torpedoes, depth charges, and a nuclear depth bomb, to be later augmented by air-to-surface missiles. This aircraft had a patrolling cruise speed of 206 knots, but could sprint at over 400 knots. The normal mission radius is 3 hours on station at 1500 feet altitude at a distance of 1350 nautical miles from its shore-based runway. This provided a total endurance of 10 to 13 hours. Available both during the Cuban Missile Crisis and after, the Sikorsky SH-3 helicopter and the Lockheed P-3 Orion were cores of U.S. airborne ASW for many years after.

These were precisely the equipment, methods, and organization the U.S. Navy used during the Cuban Missile Crisis in the fall of 1962. The U.S. had detected Soviet land-based nuclear missiles being constructed on Cuba. Additionally, Soviet submarines had been tracked while transiting from the Norwegian Sea to the Caribbean area. The U.S. response to the presence of Soviet submarines operating in the Atlantic Ocean and the Caribbean Sea resulted in what has been called the greatest antisubmarine armada ever assembled in the Atlantic. Four of these Hunter Killer Groups were at sea along with numerous other forces to implement the blockade of Cuba ordered by President Kennedy.[41]

As the history has been researched and written, the world has learned that the Soviet Union deployed four diesel "Foxtrot" long-range attack submarines with the intention of basing them at Mariel, Cuba. They were to be followed by seven "Golf" class missile boats likewise to be based in Mariel. That these Foxtrots each carried a nuclear armed torpedo is now known. The Golf class diesel-powered ballistic missile submarines each carried three R-13 missiles with a range of 370 miles. These naval forces were in addition to the tactical and strategic rocket forces already on Cuba at the time preparing launching sites for the R-12 (NATO Code Name Sandal, SS-4) and R-14 (NATO Code Name Skean, SS-5) missiles. In addition, defensive SA-2 surface-to-air missiles and Luna coastal defense missiles were installed. This Soviet buildup of forces included Il-28 medium-range bombers and Mig-21 fighter aircraft. The intermediate-range R-14 missile had a range of over 2000 miles, the medium-range R-12 a range of 1050 miles, and the Il-28 bomber a combat radius of 600 miles. All of these vehicles were equipped with nuclear weapons. They were definite threats

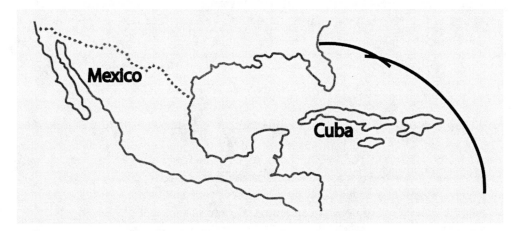

Diagram of Walnut Line arcs established during the Cuban Missile Crisis.

to much of the eastern and southeastern United States by being based on the island of Cuba, just ninety miles from Florida. The Luna coastal defense missiles, although nuclear warhead–equipped, had a range of 25 miles and were intended to repel an invasion force.

President Kennedy ordered a naval blockade of the island of Cuba. The implementation of this order resulted in the U.S. Navy establishing the Walnut Line, as a series of arcs five hundred miles from the nearest points of Cuba. At this line the United States intended to examine all cargo ships bound for Cuba and prevent passage to any ship carrying weapons or military equipment. It is along this Walnut Line that the U.S. ASW forces detected, tracked, and located the four Soviet Foxtrot submarines.

The Soviet Union perceived this confrontation we call the Cuban Missile Crisis as a colossal failure of its forces. The failure was on numerous levels. First, and even prior to the sailing of the diesel submarines, the Soviets had insufficient surface vessels to mount an effective screen of protection for the fleet of cargo ships engaged in the transport of military hardware to Cuba. Additional construction of cruiser ships had been canceled by Soviet Premier Nikita Khrushchev to focus resources on the building of nuclear submarines, which were not ready for these events. Second, the four long-range diesel submarines were in fact detected and rendered considerably less of a threat by the U.S. ASW forces. And third, the Soviets had had to withdraw the land-based missiles and other equipment from Cuba. The resulting next spiral of technological improvements to make their submarines quieter, and hence stealthy to the American detection methods, led the U.S. to likewise improve its ability to detect and hence defend against the Soviet improvements.

Although in limited use by the U.S. during the Cuban crisis, the long-range detection equipment known as SOSUS, for Sound Surveillance System, greatly expanded in the decade following the 1962 crisis.

In part the expansion of the Soviet Navy was in response to a major advance by the U.S. Navy in naval technology. In 1957 the USS *Nautilus* had initiated the true submarine type of vessel. The nuclear-powered submarine was capable of remaining below the surface for extended periods of time. It primarily needed to surface only to communicate with its command authority. The nuclear-powered submarine has been called the perfect submarine hunting platform. This claim is largely because it can operate beneath the surface and enter the same water as the opponent if necessary. This capability would remove a major tactical advantage of the adversary submarine: the ability to seek refuge in water with different acoustic properties from that of the hunters' listening equipment.

Keeping Up with the Soviets' Improving Submarines

The equipment and methods of locating a submarine had not changed drastically after the period described in the previous sections, but evolution in aircraft, electronics, and tactics had to take place. For one thing, the Soviet submarines of the 1970s and '80s were more capable that those of the '50s and '60s. The western navies had to advance if they were to locate, track, and potentially attack the newer submarines.

In the other areas of defense equipment that the western nations developed during the Cold War, for providing air and missile defense, that equipment was not replaced just because technological improvements made newer equipment more efficient. In the case of the submarine defense, technological evolution of the defense equipment was required because the Soviet Union continually evolved more capable submarines over the entire period. The following diagram depicts the years in service of first the Soviet submarines, second the U.S. Navy surface vessels, third the U.S. Navy ASW aircraft, and the electronic upgrades of the ASW aircraft.

In each series of Soviet submarine design there were generally three types: an attack submarine, a guided missile submarine, and a ballistic missile submarine. The U.S. and western navies needed to do ASW for all three types to defend convoys from the attack submarine, naval fleets from attack by guided missiles, and the continent from attack by ballistic missiles.

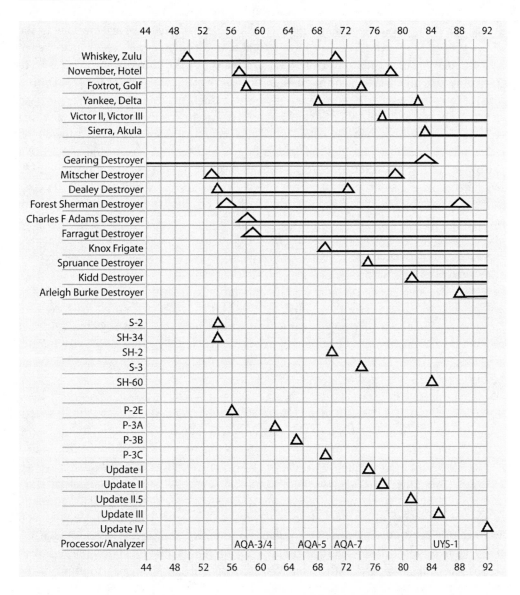

Chart of evolution of Soviet submarines and U.S. anti-submarine equipment over time.[42]

The U.S. was motivated to make these improvements because the Soviet Union's navy continued to improve its submarine fleet. Their initial postwar subs were the Whiskey and Zulu classes, in service from 1949 to 1958 and 1952 to 1971 respectively. These two types were direct applications of the technologies of the German Type XXI, which both the western Allies and the Soviets had captured and copied at the end of World War II. The fleet of Soviet subs grew to over 400 based on these designs. The Soviets first launched a ballistic missile from a Zulu submarine in 1955. The Golf class diesel-electric submarines followed the Zulu beginning in 1958, and carried three missiles with a range of 370 miles. The Foxtrot class diesel-electric attack submarines were also launched in 1958. The first Soviet nuclear-powered submarine was the November class of 1957. The November was followed by the Hotel and Echo classes, ballistic missile and guided missile types respectively; these were the so-called HEN submarines due to the similarities in the machines. They built a total of 55 of these types. The Hotel class of ballistic missile submarine performed the first patrol in the North American Basin in April of 1962. These waters are bounded on the west by the east coast of the U.S. Four of the Foxtrot attack submarines, each equipped with a single nuclear torpedo, were detected by U.S. ASW forces during the Cuban Missile Crisis in the Caribbean Sea. Several of the Golf class missile boats were to have shared the same port near Havana had the Soviet plans come to fruition.[43] In 1968 the Soviets began to deploy the Yankee class SSBN to a patrol station east of Bermuda, part of the Yankee-Charlie-Delta series of subs. By 1971, the Yankees were also on patrol stations in the Pacific. In the late 1970s the Delta class boats assumed these duties. The second generation of nuclear attack submarines were the Victor I, II, and III. The Victor III was the quietest of the series. The Typhoon class of SSBN was the final, the largest, and the most sophisticated of the missile launching submarines. The Akula and Sierra attack submarines incorporated advances based on espionage-derived findings.

With the Soviet deployment of ballistic missile launching submarines to waters adjacent to the U.S. from where they could directly launch missiles onto North American targets, the U.S. initiated a countering move. It was decided that each Soviet missile boat south of the GIUK gap would be followed by an attack submarine in addition to a P-3 aircraft tracking from overhead. Most often the aircraft shadow would be on station for four hours before being relieved by another aircraft.

This cyclic response that the U.S. needed to improve to be able to detect their improvement took a significant turn in the 1980s. Beginning in 1968 and continuing until 1981, a group of spies with access to U.S. Naval secrets supplied information to the Soviet Union. What became known in the news media as the Walker Spy Ring was made up of civilians working in the Department of the Navy. They provided the Soviets with information on how good American submarines were, and on how bad the Soviet submarines were, in the vitally important quality of sound emission by a submarine. This emission of sound is the principal attribute of a submarine that the acoustic monitoring equipment, and all the procedures described, are reliant on to detect and track a given submarine. If the submarine is quieter than the hunters' equipment is able to detect, the hunt will be unsuccessful. Throughout the Cold War, the U.S., or the West, had to improve the sensitivity of the searching equipment in order to detect and track the improved submarines of the Soviet Union.

Sensing Evolution

This period of the Cold War, during which great emphasis was placed on matters of defense, saw significant changes in the sensing technologies. The forces to provide the defense against potentially attacking submarines certainly benefited from this evolution.

In general, for a sensor to be useful it must respond to some quality exhibited by the object to be sensed. In the case of the submarine, the unique qualities that sensing could exploit to determine presence are the following:

a. Metal could be exposed above the surface of the ocean.

b. Anything exposed above the surface may be identified visually.

c. A large ferrous object could be present near the surface.

d. A sound source could be present beneath the surface.

e. Radio frequencies could be transmitted above the surface.

f. Diesel exhaust of the engines in the air.

g. If snorkeling, the heat of the diesel exhaust could be present above the surface.

h. Anything exposed above the surface may have a different thermal signature from the surrounding ocean.

The array of sensing technologies applied to sense these qualities during this period included the following:

Search Radar: A radar is a combination of transmitter and receiver. The transmitted electrical signal is reflected off some distant object and the reflected signal received by the radar receiver where it is processed for display to the operator.

These generally are used to look for objects with enough metallic content to reflect the radar energy. Targets include ground vehicles, airborne vehicles, land masses, surface ships, submarines on the surface, submarine periscopes, and submarine snorkels.

These sensors are typically mounted on a moveable arrangement, so that the direction of the transmitter and receiver can be pointed in a desired direction, and can be controlled either manually or through some automation. This arrangement allows for movement in both azimuth and elevation, from a few degrees to 360 degrees in azimuth and somewhat less in elevation. The processing and display electronics of the system incorporate features to select the portion of the total of returned information to be displayed or recorded.

Range is limited to the line-of-sight to the local horizon. A radar system provides bearing and range to a target's location.

Searchlight: Illuminates an area in darkness. Allows visual search of that area at night.

Magnetic Anomaly: Monitors and indicates minor fluctuation in the earth's magnetic field. Such changes would primarily be caused by a ship or submarine, or other large ferrous object.

Range is limited to several hundred feet.

Passive Acoustics: Receives audio signals, sounds, which can be relayed to a receiving aircraft through a radio communication link. This is the case with the sonobuoys used to locate and identify submerged submarines.

Range of the underwater audio signal is limited to several miles from the microphone or hydrophone. Range of the relayed signal is limited to the line-of-sight to the local horizon, if this communication is in the VHF band. A receiver located at the transmitting location can generate both bearing and range information to the source of the reflected sound. Receiving a single sound at three separate listening locations can be reduced to a position fix on the sound source.

Active Sonar: Receives the reflected audio signals from some distant object as a result of transmitting a sound pulse in water. Range of an active sonar is limited to a few miles. Most naval ships have active sonar equipment.

ESM: Receives wide band electronic radiation from a distant transmitter. Generally this technique is used to identify the type and location of that transmitter. The type of signal is classified by similarity to entries in a library of signal information built up over time. For radar type signals, frequency, pulse width, and pulse repetition, in addition to location are used to identify the type and use of the transmitter.

Range is limited to the line-of-sight to the local horizon. Generally this system provides a bearing to a transmitter. Multiple bearings are used to establish an area for the transmitter's location.

Exhaust Sniffer: Detects the particles present in the exhaust of a combustion engine. The particles are mostly a residual of the fuel.

Low Light Level TV: Amplifies the available light in an area in darkness. Allows visual search of that area at night. This image is presented to the operator as a television-like image on a video monitor.

Infrared Sensor: Displays thermal differences of objects. Allows visual search of that area at night and in other conditions of limited visibility. This image is presented to the operator as a television-like image on a video monitor.

The active sensors, both sonar and radar, have the drawback that they transmit a pulse of energy to be reflected from the target object. This transmission can be detected by the target and the detection range by the target is two to five times the effective detection range of the sensor. Each of these sensors provided either a line-of-bearing, or an arc at a certain range, or both so that this information could be used to determine the location of the sensed object.

These various sensor equipments could be utilized on the various platforms, as indicated in the following table.

Sensor	*Submarine*	*Destroyer*	*Helicopter*	*Aircraft*
Radar	Only used while on the surface.	Air & Surface Search	Surface Search	Air & Surface Search
Searchlight	N/A	Yes	No	Yes
Magnetic Anomaly Detector	N/A	N/A	Yes	Yes
Passive Acoustics	Towed Sonar Array	Towed Sonar Array	Passive Sonobuoys	Passive Sonobuoys
Active Sonar	Yes	Yes	Dipping Sonar	Active Sonobuoys
ESM	Only used while on the surface.	Air & Surface Search	Air & Surface Search	Air & Surface Search
Low Light TV	N/A	Yes	Yes	Yes
Infrared	N/A	Yes	Yes	Yes

Table of antisubmarine sensors applied to the various search platforms.[44]

The employment of the aircraft carrying these sensors involved creating and perfecting tactics that would make best use of the sensors available. The basic sequence of submarine location following the search-initiating event differed for the aircrew and equipment of the earlier period of equipment and the later period.

P-2/S-2 Period	*P-3/S-3 Period*
1. Gain initial contact with a Jezebel pattern, or radar observation, or visual observation.	1. Gain initial contact with a LOFAR pattern, or radar observation, or visual observation.
2. Localize with two CODAR sequences.	2. Localize with DIFAR sonobuoy.
3. Enter Julie (Active ping with underwater charge) pattern with 4000 yard or 8000 yard pattern.	4. Achieve final location with MAD detection.
4. Achieve final location with MAD detection.	

Table showing the basic sequence of actions for a submarine detection for the earlier and later periods of the Cold War.

In addition to the long-range detection of the SOSUS networks, improvements in sonobuoys and the receiving and processing equipment aboard the aircraft would also extend the coverage area for the searching and responding patrol aircraft. Whereas at the end of World War II the long-range patrol aircraft could monitor a single sonobuoy at a time by listening to the radio channel the buoy transmitted, on as the Cold War progressed, first four buoys simultaneously, then eight, then sixteen, then 32 could be monitored simultaneously by first one operator, then two. Advancements in radio receiving equipment brought the multiple channels into the aircraft, but these could not be listened to simultaneously. Along with the multiple channels, the LOFAR process described previously brought the concept of monitoring the multiple channels through a visual display that was first mechanized by the gram printer, to be followed several years later by the electronic CRT display.

Although sonobuoy development had occurred in the earlier periods, during this period a number of new sonobuoys came into being. Buoys were developed to meet specific needs, as well as to perform the basic sensing operations in an economical manner. The bathythermograph buoy was developed to allow the air crew to determine the characteristics of the water where they were operating. Similarly, a calibrated LOFAR sonobuoy was developed to allow the crew to accurately record the sounds from a particular submarine, so that the recording could be analyzed and the detailed parameters were then associated with that type of submarine.

The application of these sensors was supported by the navigation equipment and process that the crew applied to determine their position. The above sensors all provided a relative measurement of the distance or direction of the sensed object from the aircraft. Although this information could be used immediately by the crew, if communicated to others, or merely recorded to enable returning to the same spot at a later time, this relative information had to be translated into a common reference scheme; this has been for some time the geographical coordinate system of latitude and longitude.

PUTTING IT ALL TOGETHER

The airborne search process was improved by additional electronic equipment on board the aircraft. The crew's knowledge of their location was improved through the use of newer navigation devices. The older method of estimating one's position through dead reckoning, the process of keeping a record of aircraft track over the earth, speed, and time was improved upon. One instrument that improved this record was the Doppler navigator. This device used radar signals transmitted from the aircraft to measure the speed traveled over the earth. The second instrument to improve the crew's understanding of their location was the dead reckoning computer. This computer took the various inputs from the aircraft, including the Doppler navigator, to create the inputs to the Dead Reckoning Trace (DRT), a paper plot of the movement of the aircraft. Annotating this paper plot with points of significance such as where sonobuoys were dropped, where radar returns had been received, or where ECM detections had been made, greatly improved the crew's awareness of the relationship of these locations; this is called situational awareness.

Similarly the aircraft position and other significant locations were displayed on an electronic display called the ASW indicator. These instruments were central pieces of electronic equipment on the P-2 and the P-3A/B aircraft. With these instruments, the crew combined the several pieces of information from the separate sensing equipment and the various operator inputs into a single pictorial representation.

The dead reckoning computer provided ground position information from the inputs of other instruments on the aircraft. This equipment was designed primarily for ground stabilization of a radar presentation. By inserting local magnetic variation, magnetic heading supplied by the compass system was converted to true heading. True airspeed was supplied by an airspeed transducer. The resultant airspeed vector (defined as true airspeed and true heading) was resolved into N-S and E-W components in the computer. The wind information was manually inserted into the computer by the navigator, and was likewise separated into N-S and E-W components. The computer combined these individual components to generate the components of the aircraft's motion. In a ground stabilized PPI display, the sweep origin is continuously displaced at a rate corresponding to the motion of the aircraft relative to the earth's surface. Thus, displays containing fixed objects on the surface of the sea or ground could show these objects as stationary. Moving targets were shown with their true speed and direction. The dead reckoning computer also provided N-S and E-W data for the Dead Reckoning Trace and the navigation plotter instruments.[45]

The ASW Indicator group, located at the tactical coordinator station, was a multipurpose, antisubmarine warfare display indicator. It was used to receive and process position data from submarine detection devices carried in the aircraft. Processed data and a radar ground display are presented on the cathode ray tube (CRT). A radar ground display could be selected to remain stationary, or to move in a direction opposite to that of the aircraft. The sweep origin represented the aircraft position in relation to the ground display. Each of six markers when displayed appeared as a dot on the CRT and represented the instantaneous aircraft position at the time of marker activation. Each of the six markers could be switched between a slewed position and the original non-slewed position, making it possible to store 12 positions of the aircraft; any six such positions could be viewed at one time. Each marker remained fixed with respect to the radar ground display except while the marker was being manually slewed. Each of the six markers could serve as a focus of an ellipse presented on the CRT. A range circle or a bearing line could be selected to appear with either the center or end point tied to the location of certain markers. A range and bearing cursor could originate from the aircraft position. Each of four remote stations in the aircraft could activate a pre-assigned marker, allowing crew members to indicate the position of something of significance on the display. Input data for display on the indicator group was supplied by the following associated equipment: the search radar, the Jezebel sonobuoy indicator, the Julie sonobuoy receiver chart recorder, and the dead reckoning computer. The ASW indicator group, in turn, supplied an input to the bearing, distance, heading indicator (BHDI), and to the ground track plotter, both visible to the flight crew. Either or both of these indicators could be used to direct the pilot to fly to a designated point to drop a sensor or weapon or just as a relative reference point. Because of the many switches and knobs on this instrument to allow the operator to control all of these features, the ASW Indicator group was nicknamed the Chinese Television Set by the crews.[46]

In the P-2 and P-3A aircraft, the Tactical Coordinator, or TACCO, used the ASW indicator display to build up a representation of the tactical situation of the aircraft, plotting the position of the sonobuoys and/or other points of interest. By using these drawing features, the operator could identify graphically the intersection of multiple symbols drawn on the screen. For example, the intersections of two ellipses, the intersection of two range circles, the intersection of a bearing line and any of another bearing line, a range circle, or an ellipse, could all be used to identify a point of interest. Another marker could be designated at one of these intersections. This pictorial representation of the aircraft's tactical situation is called the tactical plot.

As with the other topics in this book, the ability to do antisubmarine warfare went through many changes during the forty-plus years of the Cold War. Airplanes and ships became faster and could operate farther from their bases for longer periods of time. But it was the technological evolution of particularly the sensing equipment that made the most dramatic changes, and the most difference in the resulting defense.

The P-3 aircraft was adopted by the U.S. Navy for the ASW role in 1962; it performed these duties during the Cuban Missile Crisis. It still serves in this capacity today, although the equipment it carries has evolved significantly. The airplane itself went through its own evolution. The original was the P-3A, adapted from the Lockheed Electra passenger airliner. This was followed in 1965 by the P-3B, which had more powerful engines and was strengthened in specific areas. To a degree, the P-3B was an interim upgrade until the more capable electronic system of the P-3C became available in 1969. The P-3A electronics initially was quite similar to the final P-2 equipment. The notable exception was moving to the inertial navigation system as the primary instrument to provide the position information as reference to all navigation operations.[47]

A fundamentally different concept for the electronic system on board the submarine hunting aircraft entered development in 1960. This concept was known as A-NEW, for ASW New, early in its development. The most fundamental change to be incorporated was the central storage and processing of a digital computer. This required all of the instruments that provided information to the system to become digital. For example, in the later P-2 and the P-3A aircraft, the navigation instruments provided their information to the central system as analog values, i.e., voltages, to be combined with related values prior to becoming the position of the aircraft displayed to the crew. In the digital system, these same values were transferred as digital numeric values which could be stored in the computer's memory.[48]

Although the set of features of the P-2 and P-3A/B electronics was different from the intended replacement primarily in flexibility and capacity, the older equipment would have been a severe limitation for the newer envisioned operations. The objective of the upgraded installation was to enable the crew to monitor fields containing many more sonobuoys, and to more quickly identify the position of the target submarine. This would only be practical if the storage, computation, and display of geographic points was very flexible. Additionally, a prime advantage of the proposed digital system was that a single set of stored information could be accessed for different reasons by the different crewmen. This allowed for expanded use of displays and key sets at the different crew stations for improved situational awareness and ease of inputting information. The resulting operation relied much less on voice communication between crew members to relay specific details of the situation to be brought to the attention of the TACCO for incorporation into the "tactical plot."

Accordingly the crew complement of the P-3C evolved with these changes. Going forward the crew would be made up of pilot, co-pilot, and flight engineer to operate the aircraft. The Nav/Com operated the navigation and communication equipment. The tactical coordinator (TACCO) both operated the computer station to create and manage the tactical situation display of the ASW operation, and supervised the sensor system operators. The Sensor 1 and Sensor 2 acoustic operator positions had identical acoustic processing equipment and shared controls of common equipment. The Sensor 3 non-acoustic operator position controlled the radar, magnetic anomaly, and ESM instruments. Each of these sensor operators would send information from the equipment at their stations to the TACCO for incorporation into the common representation of the assigned problem. The in-flight technician monitored the on-board electronics and could resolve some equipment malfunctions that otherwise might terminate a flight. The ordnance technician was responsible for monitoring the search stores on the aircraft, primarily the sonobuoys, and loading those sonobuoys stored inside the aircraft into the appropriate launch chute at the direction of the TACCO.

The TACCO viewed the tactical situation on the screen in front of him as he worked with it. The pilot and co-pilot had a display screen on the flight deck positioned between them that displayed at least a portion of the same information that the TACCO could see on the larger screen of the ASA-70 display. The non-acoustic operator likewise had an ASA-70 display at the station, which also displayed the radar image. As part of the Update I improvements, in 1975, the same type of display as for the pilots was added between the two acoustic operator stations. With this common graphical depiction of the location of the points of interest to the ASW situation, all these crew members were better aware of the situation and better able to focus on their next task.

Another fundamental change enabled by the P-3C's digital architecture was the means to communicate the information stored in the onboard computer to other platforms. This feature was principally used in three ways. First, the P-3 could copy its information to surface ships outfitted with the appropriate equipment. Second, it could copy its information to an arriving P-3C aircraft, such as when an aircraft got low on fuel and a relief aircraft would take over the ASW problem underway. Thirdly, this communication equipment could also be used to receive information in the aircraft. This allowed the crew to receive updated information after departing base. All three scenarios enabled multiple platforms in the vicinity of and working the same search or investigation to have the latest information. This information included the type, status, and location of the sonobuoys in the water and other points of interest such as the location of radar or ESM observations. Similarly, the

P-3C would begin a mission by downloading information from a magnetic tape provided by the support facility for that unit. Mission information could be printed on the on-board teleprinter as needed.

With the aircraft's ability to monitor more sonobuoys simultaneously came the need to record more channels of the raw sound information from the sonobuoys. These audio recordings were part of the record of the crew's work, and could be reviewed following a flight as a double check of the flight. Where the final P-2 had two channels of audio recording, and the P-3A/B had fourteen channels, the P-3C increased this number to 28. This enabled greatly increased review of the inflight decisions made by the crew. Postflight review of these tapes could produce a submarine detection that had been missed during the flight. This review was conducted by the shore-based Anti-Submarine Warfare Operations Center (ASWOC) or the Tactical Support Center (TSC). The name of these shore-based support facilities changed over time, but their responsibilities overlapped greatly; they initiated each actual ASW airborne mission, supplied the information the crew would begin the mission with such as maps and conditions in the mission area, provided the computer tapes to make current the information stored in the onboard computer, and received and handled the information the returning crew had collected during a mission.

The P-3C developments removed the diesel sniffer and the searchlight from the set of sensors capable of detecting the presence of a submarine. Although the diesel exhaust of the snorkeling submarine was still evidence of its presence, this sensor had been less than a complete success due to the presence of similar exhaust from the plethora of commercial vessels on the seas, and the difficulty the crews had in positively tracing the sensor indication to the submarine. Also, modern patrol aircraft have deleted the searchlight in favor of low light television (LLTV) or infrared (IR) cameras, or perhaps both. The objective is the same: locate the contact in darkness so that precise location can be determined, followed by identification.

Among the airborne sensors applied to the ASW problem, it was the instruments associated with the passive sonar that saw the most marked change over time. The sonobuoys dropped into the water underwent evolution and performance improvements to improve the detection, localization, and tracking operations. And the processing system on board the listening aircraft saw significant improvement.

The four main areas for improvement were in the number of sonobuoys that could be monitored by a single airplane, the ability to process the signals from more than one type of sonobuoy, the ability of the operator to identify and focus on the signal of interest, and the speed at which this information could be used to establish a position of a submarine. Evolving to equipment that automated some or all of the tasks performed by the operators was crucial. In earlier days, the operator had to devote a large amount of time to repetitive steps in the operation of equipment to bring out the needed information. This focus on adjustment of equipment prevented the acoustic operator particularly from assessing the tactical situation, such as establishing an expectation of the received signals.

The U.S. had initially not had good results with directional sonobuoys, chiefly due to low reliability. As a result, other methods were developed to establish the direction to a sound source with the omnidirectional buoys available. One method was the previously described Julie procedure to determine the location of a sound reflector. Also previously described was the CODAR process to determine a line-of-bearing to a detected submarine.

These approaches to determine the direction from the source of sounds were initially improvements to the LOFAR only detection, which provided no directional information. This step was part of the localization task following detection of a submarine. The CODAR process, however, had drawbacks. It took time for the airplane to fly the pattern to properly place the four buoys. After placement of the buoys, the operator would need to monitor the signals for a time to allow a trend to be established in the display. During this entire time the aircraft was operating at low altitude fairly close to the adversary submarine. The fact that the CODAR pattern took four buoys to determine this angle provided an expense which could be retasked into a different sonobuoy. The drawback of the Julie process was that the sound source dropped by the aircraft advertised the presence of the searching aircraft nearby. These limitations resulted in development of the DIrectional Frequency Analysis and Recording (DIFAR) sonobuoy. This buoy transmitted both an omnidirectional hydrophone signal, compass information, and two hydrophone signals oriented a fixed angle apart called the N/S and E/W signals. The more complex signal processor required for this buoy used the two directional signals to determine the angle of arrival of the sound. This more capable signal processor was initially the AQA-7 of the P-3C. So although this DIFAR buoy was more expensive than the simpler omni buoy, it offered the advantage that a single buoy could both provide the initial detection of a sound source, and an indication of direction to that source.[49]

These improvements to the ability of the listening equipment occurred in several steps. Improvements in sensitivity to smaller signals occurred with the introduction of the AQA-5 in 1966, the AQA-7 in 1969, and the UYS-1 in 1985. Similar to the improvement to the operator efficiency of the acoustic processing equipment, the ECM system on the P-3C performed many of the operations described earlier automatically, freeing the non-acoustic operator of many low level tasks of ECM equipment operation. These and other improvements to the onboard electronics system were part of Updates I, II, II.5, and III of the P-3C, implemented between 1975 and 1985.[50]

Visual search from either helicopter or fixed-wing aircraft were both optically and electronically enhanced with the arrival of the imaging turret on the scene. This system held one or two electronic cameras, the optics needed by the cameras, a gimbaled turret, and supporting electronics. The gimbaled turret allowed the operator to point the camera in a desired direction by operating control switches mounted on a joystick. The turret could also be directed to point by a command from the aircraft's computer. Outputs of the instrument were the camera image and the angles that the turret was pointing, as with the other sensors providing a relative position to an object in view. The operator viewed the output of the TV-like camera on a video monitor. The cameras were a selection from among the types available: daylight (TV), infrared (IR), or Low Light Level Television (LLLTV). The daylight camera would be used during daytime and would provide enhanced visual search for the crew. The infrared camera would be used at night or other times of reduced visibility. The LLLTV camera amplified an available light source and also was primarily used at night. These cameras used in darkness were especially useful in the ASW search to detect the exposed snorkel of the diesel submarine at night. All of these cameras provided a video output that could be recorded for additional review later. On the earlier aircraft installations, film cameras had been mounted in the belly of the airplane to photograph objects of interest. These photos were only available following film development on the ground and so were not a tool for the crew to use during flight. This imaging turret instrument was included in the equipment of the S-3 aircraft when introduced in 1974, and was added to the P-3C in the Update II of 1977.[51]

To maintain an up-to-date location of sonobuoys in the water, prior to the P-3C the crew had to periodically fly over the top of each sonobuoy and mark the latest location; the "Mark-on-Top" maneuver. This entailed the pilot's maneuvering the airplane to each buoy by homing on the radio transmitter of the buoy, and hitting the on-top button so that the TACCO could update the marker representing that buoy. This had been another set of tasks the crew had needed to perform to keep the location of points current to account for the effects of ocean currents and winds on the position of the buoys. It was also another chore pulling the crew away from their principal task of finding and classifying the submarine. An instrument added to the P-3C in the Update II would remove much of this workload. The Sonobuoy Reference System performed angle measurements on the signals arriving from a set of sonobuoys, through a set of antennas with very specific geometry. The system enabled updating of the computer's location for the various sonobuoys without crew interaction. This periodic updating of the system's representation of the sonobuoys' locations, offsetting the effect of ocean currents, for instance, improved the estimates of locations relative to those sonobuoys.

These improvements made each aircraft and crew more capable. They could monitor a larger area, search for quieter submarines, and provide means of detection not previously available.

Mission Refinements

In order for submarines of the Soviet Union to depart their home waters to take up positions from where they could threaten the American mainland, they had to pass through fairly narrow gaps between land masses. In the Pacific this was north of the Japanese islands, and in the Atlantic this was one of the passages of the GIUK gap. It was in these areas where the submarine hunters would use a type of barrier to listen for the transiting submarine. As nuclear-powered submarines replaced the diesels, the sub's ability to remain submerged for long periods of time made the acoustic sensing the primary method of detection. Similarly, since it was the objective of the U.S. Navy to constantly track all Soviet ballistic missile submarines, these patterns of sonobuoys were used if it was necessary to relocate a submarine. The increased importance of these choke-point tactics due to the realities of global geography led to additional evolution in the use of sonobuoys.

A pattern of sonobuoys was established in order to accomplish a specific task with respect to a hostile submarine. One of the following missions would be assigned to a particular aircraft, and that aircraft would drop the corresponding series of sonobuoys and monitor their signals. In the following diagrams, the circle drawn around the central location of the buoy represents the detection range of the buoy. Notice that the circles overlap and there are no areas within the pattern which would not generate detection from at least one buoy. For example, if the detection range of the buoys is 7.5 nm, the center-to-center spacing of the buoy pattern is 10 nm.

Area Search

The objective of this pattern, known as a distributed field, is to cover an area and generate at least one contact from a submarine passing through or about to pass through this region.

Barrier Line

The objective of this pattern is to generate a contact or warning when a submarine crosses the line. There are two main reasons to establish such a line: first, when the line is established at a distance from an object of interest, say a friendly high-value ship or group of ships, the warning generated by the contact alerts the defensive force that a submarine is approaching. Second, when the line is established near a harbor, naval base, or other coastal entry/exit point, the warning generated by the contact alerts the force

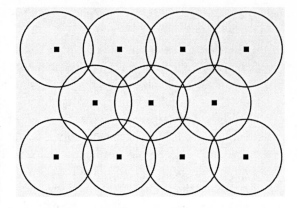

Diagram of the sonobuoy pattern used for an Area Search.

that a submarine may have left that area. In either case, one of the other patterns is usually established soon to move the detection to the next phase of the ASW process.

When the submarine enters one or more of the detection circles it is detected, giving an initial position, termed datum, for its location. The arrangement of buoys depicted below shows the detection range of six buoys overlapping; this results in a detection probability of 100 percent. A target crossing the line of buoys as depicted would be detected. This may be required in certain circumstances, but also may require more buoys than is economical under other circumstances. A lower probability of detection may be the sacrifice for lengthening the line to cover a wider swath of ocean with the same number of buoys.

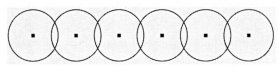

Diagram of the sonobuoy pattern used for a Barrier Line.

The barrier line of sonobuoys would also be used by the ASW aircraft of the carrier battle group to provide a warning line along both sides of the moving carrier battle group formation. As the group moved along, a new sonobuoy would have to be added to each barrier line to provide the warning protection as the group of ships sailed ahead. If these defensive barriers detected an unknown submarine, either the battle group's fixed-wing or helicopter aircraft, or both, would investigate the contact, likely dropping additional sonobuoys and/or employing their other sensors.

Walking Barrier

The objective of this pattern is to screen a large area by constructing a group of barrier lines, in effect marching the line of buoys in a desired direction. Each barrier line is monitored for a time, perhaps 2 to 3 hours, before laying the next successive set.

Datum Prosecution

Investigating an area around where the submarine was last thought to be was quite similar to that described in a previous section. The distance between buoys could be greater due to the improved equipment.

Handoff

Although not strictly a search pattern, a set of buoys could be used to mark the location and direction of a submarine of interest at the time. When an aircraft had to depart the area of a tracking operation, it could leave a sonobuoy at the location of the submarine just prior to departure, and then leave another sonobuoy several miles away in the direction of the track of the submarine at that time. This would be a physical message to an aircraft arriving in the area. More often, the departing and arriving aircraft crews would use voice communication to coordinate so that neither aircraft would be in danger of another aircraft operating in the area.

Surface Ship Evolution

Evolution was also required in the sensors and weapons of the ASW surface ships, or destroyers; some types would later be called frigates.

At the end of World War II, the ASW weapons in use by U.S. ships were depth charges, the homing torpedo, and deck guns. Ships with this arrangement remained through multiple steps of evolution in the ships employed in ASW. As the range of the weapons available for a submarine to attack the high-value aircraft carrier at the center of the fleet improved, the distance the defensive screening ships were stationed from the carrier increased. Similarly, the range at which the destroyers of the defensive screen needed to detect and attack an approaching intruder increased accordingly. This led to more capable sonars, radars, and weapons for the surface ships. As illustrated previously, traditional depth charge attacks required the attacking ship to be essentially on top of the submarine. This became impractical as the Soviet subs were able to launch missiles at the principal assets of the fleet from some distance. Defense of the fleet required being able to attack the adversary submarine at least at the distance he could attack from.

An incremental improvement was the spigot mortar, named Hedgehog, which threw a dozen charges 250 yards in front of the ship to create a pattern of charges in the water. The next improvement was Weapon Alpha, which fired up to 22 rocket-propelled depth charges from 400 to 800 yards from the ship. This was followed by weapons that used rockets to transport a weapon out some distance from the ship, where the weapon would then perform similar to the previous generation. Both depth charges and homing torpedoes had a later counterpart which could be launched to a point five miles from the ship. The Anti-Submarine Rocket, or ASROC, launched either a homing torpedo or a nuclear depth bomb from 1000 to 10,000 yards from the launching ship. This ASROC was often collocated on a ship equipped with a Drone Anti-Submarine Helicopter (DASH), an unmanned helicopter that could carry a torpedo some 22 miles from the ship. These weapons provided the ship a standoff range to attack the submarine, and allowed for another unit of the fleet to detect the submarine and relay information to a ship to launch these longer-range weapons.[52]

A variant of shipborne sonar sensor that was evolving through this period was the towed sonar. This itself had two types: the variable depth sonar and the towed array. For both cases a cable was suspended behind a moving ship, normally a destroyer or frigate, and the sensor was thus in the water some distance behind the ship. The main component of the variable depth sonar was the "fish" that was towed; it had controllable surfaces that allowed it to be steered to a deeper or shallower depth, depending on water conditions.

This fish could provide either active or passive sonar to cover the vicinity of the ship. The towed array contained a number of sensors distributed along the cable, normally more than a mile in length, and the principles of its operation were quite similar to a SOSUS array of hydrophones, previously described. The towed array would primarily sense the areas perpendicular to the essentially straight line of the towed line. For both of these sensors towed behind ships, a key concept was that by placing the sensor element some distance from the ship, that ship's noise would interfere less with the sensing operation. A ship using the towed array would normally be positioned on the outermost periphery of the battle group, searching outward to detect an unknown hostile. The U.S. submarines would also make heavy use of the towed array sonar following its inception.[53]

Throughout the early Cold War years the destroyers of the U.S. fleet were of types designed and built for World War II. With the improvements in the Soviet navy, the sensors and weapons of the U.S. ships were not adequate. The option of replacing them with new ships was not realistic; there were too many of them and there were other federal spending priorities. What was decided upon was a set of Fleet Rehabilitation and Modernization (FRAM) programs to alter the primary mission of these ships from attacking surface ships to submarine hunting. While these modifications covered much ship equipment, the key changes were to upgrade the sonar to that of greater range, and upgrade the weapons to a combination of the weapons described above. With these modifications completed between the years 1959 and 1965, these ships continued to serve into the 1980s. The oldest of the ships converted saw removal of some of their guns to be replaced by the Hedgehog launchers. Another group of ships, with sufficient room, received the Weapon Alpha rocket launcher, newer torpedo tubes, and the hangar and flight deck for the DASH helicopter. Another group of ships received the ASROC launcher, the newer torpedo tubes, and the DASH helicopter capabilities. Most of the converted ships received the SQS-23 sonar, which provided sensing out to 10,000 yards (6 miles), to complement the longer-range weapons. A number of these converted ships were also outfitted with a variable depth sonar (VDS). Not directly related to the ASW mission, these ships also included upgrades of antiaircraft weapons, with the earlier guns being replaced by rocket-propelled guided missiles. A total of 170 ships were planned to be converted with these changes.

Beginning in 1969, the ASW helicopters were moved from the central carrier platform to the ships distributed around the periphery of the carrier battle group, as newer ship designs entered the fleet. This not only freed up room on the carrier for other types of aircraft, but it placed the helicopter closer to the source of a threat submarine. Knox class frigates, Spruance class destroyers, and Oliver Hazard Perry class frigates began to replace the FRAM upgraded ships in 1969, 1975, and 1977 respectively. These newer types entered service with the ASW weapons described for the later FRAM conversions.

The Knox class initiated the newer ASW configuration being fielded in a new-built ship: longer range main sonar, variable depth sonar, ASROC launcher, and DASH helicopter to deliver a torpedo out some distance from the ship. In the 1970s, these ships were in turn upgraded with the other newer submarine sensor, the towed array sonar, to provide the ship with a longer-range subsurface detection. These sensors had had initial trials on destroyers in the late 1960s, but had evolved primarily as a sensor aboard submarines. On either an attack or ballistic missile submarine, the towed array provided not only a long-range sensor but one that could observe the water behind the sub. Similar to the VDS on the destroyers, or frigates, the towed array could be lowered to a variety of depths to best deal with water conditions. In the upgraded Knox, this sensor was combined with LAMPS, or Light Airborne Multi-Purpose System, a manned helicopter equipped with both ASW sensors and weapons to respond to a detection made by the longer-range towed array sonar. The helicopter, initially the Kaman Seasprite SH-2, was considered an extension of the ship in that information from its sensors was up-linked to the ship where the radar and acoustic operators resided. Beginning in 1975, the Spruance class of destroyers were launched with somewhat upgraded sensors and weapons, but substantially for the same work as part of the carrier battle group.[54]

Improved Naval Fleet to Find Submarines

The next change in the evolution of the U.S. naval fleet was that of combining the antisubmarine task group and the attack task group into a single entity, the carrier battle group, to perform all the previous missions.

The carrier battle group was built around the Enterprise class of nuclear-powered aircraft carrier. These larger carriers were able to house the five squadrons of aircraft needed. Typically, the carrier would be home to air defense fighters, aircraft capable of attacking sea-based or shore-based targets, radar aircraft providing detection of both air and surface contacts, as well as the carrier-based ASW aircraft. The ASW helicopters were housed on the frigates and the destroyers making up the defensive perimeter of the battle group fleet, as the newer ship designs entered the fleet.

These platforms, sensor equipment, and tactics were combined and implemented in the carrier battle group. The aircraft carrier, located in the center of the collection of ships as the high-value asset, was home to the fixed-wing aircraft. These aircraft included:

a. Fighter interceptors to respond to any incoming air threat to the force and to provide protection to the other aircraft on their respective missions.

b. Fighter attack aircraft to allow an air attack by the carrier group on either surface or shore-based targets.

c. Airborne radar platforms to provide air search over the region of the carrier group, and to provide control over either air-intercept platforms as well as antisubmarine aircraft to coordinate their movement.

d. Antisubmarine fixed-wing aircraft to place and monitor sonobuoy sensors and to respond to any reported submarine sighting or other detection.

e. Antisubmarine helicopter aircraft, also to place and monitor sonobuoy sensors and to respond to any reported submarine sighting or other detection.

f. Electronic Warfare aircraft to provide protection from adversary electronic equipment primarily to the fighter attack aircraft of the group.

The Grumman E-2 Hawkeye radar plane operates at some distance from the task force, using its radar to continually search the ocean surface for signs of a submarine, and of equal importance to search the skies for intruder aircraft. The E-2 plane, which had evolved as an improvement to the E-1 Tracer, had significantly more capable radar, and was powered by two turboprop engines. The radar antenna was contained in a 24 foot rotating disk mounted above the high wing of the aircraft. The plane has a five-person crew: pilot, co-pilot, combat information center officer, air control officer, and radar operator. The latter three operate workstations in the rear of the aircraft, to run the airborne radar site. The E-2 aircraft and its systems saw multiple upgrades between its introduction in 1964 and the end of the Cold War. A later variant is still in service today.

The Lockheed Viking was the primary antisubmarine long-range detection and local-ization system assigned to the carrier battle group. The S-3 Viking antisubmarine aircraft was powered by two turbofan engines and carried a crew of four, made up of pilot, copilot, TACCO, and the SENSO system operators. The copilot operated the non-acoustic sensors: the radar, ESM, MAD, and the forward-looking infrared sensor. The operational scenario of the S-3 was to maintain lateral sonobuoy barriers on either side of the battle group and investigate any report of potential submarine activity within striking distance of the group. To do this it could remain on station in a contact area for 7.5 hours. Its cruising speed was 350 knots with a range of 2765 miles. This time on station and range could be extended by air-to-air refueling. S-3 aircraft contained the systems for search, detection, and kill. It had an internal bomb bay that could carry a combination of weapons including homing tor-pedoes, depth charges, and a nuclear depth bomb. Antisubmarine sensor equipment included up to 60 sonobuoys launched from chutes located in the bottom of the fuselage between the wings. The on-board sensors of this aircraft were search radar, electronic sup-port measures (ESM) detection set, infrared detection camera housed in a chin-mounted turret, and magnetic anomaly detector or MAD probe. The S-3 airplane was introduced in 1974 and served into the 1990s, beyond the end of the Cold War.

The Sikorsky SH-3 Sea King antisubmarine helicopter also carried a crew of four: pilot, copilot, radar and MAD operator, and acoustic operator. The SH-3 turbo-shaft engine powered helicopter had a range of 550 nm and a speed of 140 knots. This helicopter carried an active dipping sonar and could deploy sonobuoys. The SH-3 Sea King was introduced in 1961 and served into the 1990s, also past the end of the Cold War.

Additionally, the Kaman SH-2 Seasprite antisubmarine helicopter also carried a crew of four: pilot, copilot, tactical coordinator, and sensor operator. SH-2 turbo-shaft engine powered helicopters had a range of 580 nm and a speed of 141 knots. This helicopter carried similar ASW equipment as the Sikorsky Sea King. The SH-2 was somewhat smaller and lighter and could operate from surface ships that did not have the room for the larger hel-icopters. The ASW version of this helicopter was introduced in 1973 and served into the 1990s, again past the end of the Cold War.

The final antisubmarine helicopter of the Cold War period was the Sikorsky SH-60 Sea Hawk. The Sea Hawk carried a crew of three: pilot, copilot, and sensor system operator. Sensors are the search radar, sonobuoys, MAD bird, ESM, and forward-looking infrared. This twin turbo-shaft engine powered helicopter had a range of 450 nm and a speed of 140 knots. The Navy version of this helicopter was over 80 percent common with the Army's H-60 utility helicopter. The ASW version of this helicopter was introduced in 1984 and served past the end of the Cold War. Later variants are still in service today.

The long-range land-based patrol aircraft of this era was the Lockheed P-3C, as described previously.

As mentioned above, the greater range and speed of the airborne response platforms were needed to provide response protection to the vast areas of both North Atlantic and

North Pacific Oceans to the longer-range missile attack capability inherent within the submarines of the Soviet Union. However, it was the improvements in quietness of the submarines fielded by the Soviet Union that kept the U.S. and other Western countries improving the equipment and techniques of detecting submarines.

The carrier battle group fleet configuration consisted of two types of surface ships, plus two nuclear attack submarines, in addition to the carrier itself. The Fast Frigate (FF) is an ASW ship, essentially a destroyer but labeled frigate in the ship designation changes of 1975. The FF ship, initially of the Knox class, was equipped to defend the task group by sensing out and attacking submarines. The Guided Missile Frigate (FFG), initially of the Oliver Hazard Perry class, whose function had also previously been labeled destroyer, defended the task group primarily from air attack with its surface-to-air missiles. The attack submarines were initially of the Skipjack class (SSN) of nuclear attack submarines.

The lead elements of the task group were the two attack submarines, which were positioned to either side of the leading frigate. This allowed the subs to search generally ahead of the other ships, clear of any noise the group generated, but able to communicate through the lead frigate. The fast frigate in the forward section of the formation was tasked with passive sonar search, LAMPS helicopter response to underwater contacts, and communication with the two submarines of the formation. The fast frigates positioned on either side of and at the rear of the formation likewise conducted their own passive sonar search, generally to the side of the formation. Similarly they could dispatch their embarked LAMPS helicopters to

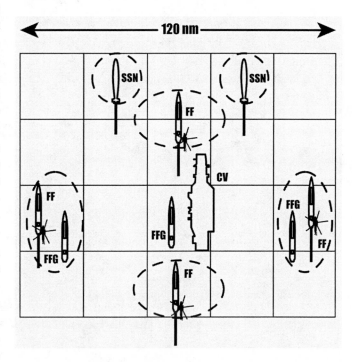

Depiction of the dispersion of support ships around the central aircraft carrier.

investigate any contact. All of these Fast Frigates and the two attack submarines were equipped with towed array sonar to conduct these passive searches, and could maneuver within their zone to improve their detection. These ships all maintain forward speed of advance of the larger group by a combination of sprinting ahead then drifting to improve sonar search as needed. To augment detection of a hostile submarine entering from the flank of the group, the fixed-wing aircraft of the carrier place and monitor a barrier of sonobuoys along the flank of the group. As the group moves along, this barrier of sonobuoys must be added to periodically to maintain the protective barrier. This barrier is placed at the distance such that a hostile sub is likely detected at a distance greater than the range of his weapon.[55]

This arrangement persisted through to the end of the Cold War, although further refinements were planned. Another significant improvement in acoustic processing was under development at the time the Cold War ended. The Update IV to the P-3C onboard system was well along, with one aircraft equipped and undergoing testing, when this project was terminated in 1992. With the end of the Cold War, these vast sums of money to change out the equipment in the numerous P-3s could be better spent elsewhere. These continued advancements were, as before, due to continued developments by the Soviet Union right up to the end of that country.

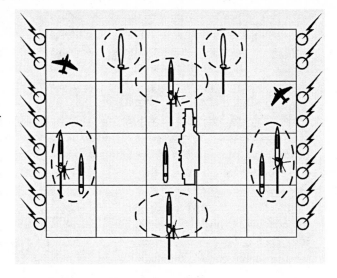

Depiction of the dispersion of the fleet with lateral sonobuoy barriers and ASW aircraft.

Chapter 4

The U.S. Establishes
Strategic Deterrence

In the previous chapters I described and illustrated the defense measures that had been established to respond to hostile aircraft and submarines that might approach the American homeland. The defenses for these threats were applicable to the physical presence of such a hostile vehicle. Although the existence of this defensive capability may have threatened the damage of an attacking nation's force of aircraft or submarines, it did not threaten that nation with widespread damage sufficient to deter the likelihood of attack in the first place. Nor did this threat of destruction to an attacking vehicle necessarily prevent the attacking nation from mounting similar attacks with additional copies of the same or similar platforms.

The next step in the evolving set of defenses was to establish deterrence to the initiation of essentially any attack. A primary shortcoming of the described defenses is that even after detecting, classifying, and tracking a threatening hostile delivery vehicle, there was still significant probability that the vehicle could not be destroyed before it launched its weapons. This was to be all the more true after the invention of stand-off weapons in the 1950s, with which the submarine or aircraft need only approach close enough to the target to launch a guided missile. This defense shortcoming was compounded by the fact, according to the practice of the nations of the world, that a potentially hostile platform located outside recognized territorial limits was perfectly legitimate. What was needed was a scheme to instill in the prospective attacker the idea that the attack itself would initiate a response, regardless of the success of the attack; and that, due to the massive damage that could result from nuclear weapons, a nation under attack needed to retaliate before the actual destruction occurred. And furthermore, because the defending country could likely not discern the specific intention or extent of a detected attack, the response would tend toward the maximum that the defender could make.

This deterrent capability has most often been an offensive force, which could be directed at the would-be attacker, capable of delivering overwhelming force of sufficient magnitude to cause unrecoverable damage to an attacking nation.

History of Aerial Bombardment in U.S. Forces

During World War I the military use of the airplane had evolved from initially that of aerial scout, to include aerial photographic reconnaissance, aerial warfare for air supremacy, ground attack in direct support of ground units, and ground attack behind the battle lines. Although all the major powers developed what came to be known as strategic bomber units, their use had been significantly limited by war's end. For example, the British RAF had developed a larger, longer-range plane capable of reaching Berlin, the Handley Page V/1500, and were preparing for a strategic bombing mission on the morning of Armistice Day. By nightfall, hostilities had ended.

Strategic bombing is directed at targets the destruction of which reduces the enemy's ability or desire to produce military equipment and forces with which future attacks could be made or sustained. Strategic targets are certainly munitions factories, airfields, port facilities, rail and highway transportation facilities, and communication facilities. But the strategic objectives may also include the people of the enemy country. Attacks on the population of the enemy are damaging in two ways: first, the workforce needed to produce the materials for war is reduced, presumably limiting the production capacity; and secondly, the will of the population both to produce war material and to fight the war directly is reduced. Conversely, tactical bombing is defined as directed at targets of immediate military value, such as combat forces, military installations, and military equipment. This definition would include destruction of items that an opposing force could use in support of making an attack on friendly forces, such as a bridge, rail line, or airfield.

Following World War I the Italian Giulio Douhet, Hugh Trenchard of Britain, and American Billy Mitchell separately proposed the proper use of the airplane in warfare. Douhet expounded the early and decisive aerial attack on the vital industrial and population centers of the opponent country. As a result, he contended, the morale of the populace would be affected so as to pressure the political leaders to end the war. Trenchard advocated that the role of an air force was to attack the enemy, and though in favor of air support for ground forces, he considered such an attack to be on the industrial and transportation capacity of the enemy nation. Mitchell prophesied that in future wars, the aircraft would become the principle and decisive weapon.[1] His concern that the imbalance in priority of funds devoted to building battleships in the 1920s would leave the U.S. vulnerable without an adequate air force. This led him to organize a demonstration that bomber aircraft could quickly destroy battleships, using captured German battleships for the demonstration. Although World War II and history proved him correct, his approach of challenging the correctness of his military superiors led to his demotion and ultimately limited his influence.

These early theories influenced planning and preparations for the next war, less than two decades away, which is now referred to as World War II. The resulting concepts of the "knockout blow" and "the bomber will always get through" greatly influenced that period and largely survive today. The knockout blow delivered by aerial bombardment is a concept that originates from the World War I experience in that the bomber was seen as being able to bypass all of the obstacles which led to the protracted and bloody ground war in World War I France. The coastline and fortifications, rivers, mountains, and long distances could all be bypassed by the bomber, and it could deliver the knockout blow to the enemy country behind these obstacles, thus ensuring a quick end to a conflict. This concept of bypassing all obstacles to attack the heart of the adversary nation was also seen as attacking the nation from within, as opposed to the traditional attack on the nation's military, which would be positioned between the heart and capital of the nation and the attacking force attempting to move inward.

With the lack of practical bombing experience by the U.S. in World War I, this newer form of warfare evolved in an ideal fashion through the 1930s. International efforts to ban aerial bombardment following World War I failed to garner sufficient support, as the major nations continued to build air forces. In the later 1930s, as events in Europe and Asia indicated that war was likely in the future, U.S. President Franklin D. Roosevelt grasped the ideals adopted and advanced by the new U.S. Army Air Corps' (USAAC) Tactical School: principally that precision strikes on critical points of a nation's industry would render that nation impotent to conduct further war, requiring them to seek peace. These critical points are those elements without which the economy, especially the making of the implements of war, would stifle.[2] Examples would almost surely include the electrical generation and distribution facilities, and the fuel production and distribution capacities. During this ideal time, it was thought that invasion by ground forces could thus be avoided.

Roosevelt significantly noted that Germany's capacity for bombardment doubtless influenced Britain's Neville Chamberlain to agree to an undesirable "peace accord" with the Nazi regime in 1938. Roosevelt pursued a large building program of bombers to establish either the capability or the deterrent threat against the aggressive nations of Germany in Europe and Japan in the Pacific.[3]

Despite the various efforts to avoid another war, it came with the Nazi invasion of Poland. Following the Japanese air attack on Hawaii's Pearl Harbor in December 1941, the U.S. was involved with wars in both Europe and the Pacific.

At the start of World War II the predominant attack element of a naval force was the battleship. These battleships, with their large-caliber mounted artillery pieces, were able to mount an attack along the coastline of an adversary, although they had been most often justified as mobile extensions of the previous shore-based defensive installations of guns protecting the nation along the seacoast. In the opening periods of the war, battleships of all the major belligerents were destroyed or seriously damaged by aerial bombing attacks. First the Italian fleet, then the German, then the American, followed by the British, and finally the Japanese all were seriously damaged by air-delivered bombs. This removed pre-war notions that a properly equipped and operated battleship was invulnerable to the aircraft.[4] This hard lesson placed the relatively new aircraft carrier as the principle source of attack capability for the naval force.

Initially, the fledgling strategic air bombardment that had continued to evolve following World War I was seen as an available method of bringing war to both Germany and Japan while the ground and naval forces needed for traditional invasion of the enemy homelands were prepared.

Two key elements of USAAC philosophy were that high-altitude daylight precision bombing could destroy the key infrastructure of a nation's economy, and that bombers defended by guns could penetrate to an enemy target area without escorting fighter planes. The first relied heavily on use of the Norden bombsight instrument. The second relied heavily on specific placement of aircraft within a larger formation to optimize the fields of fire of the bomber-mounted guns.

At the start of World War II, the U.S. had minimal standing air forces. The small number of regular officers were spread across many newly established air groups as a small foundation of professional military men. Lt. Curtis E. LeMay quickly became Col. LeMay, commander of the 305th Bomb Group, supported by numerous 90-day wonders. LeMay's job, and that of his contemporaries, was to prepare his group, mostly comprised of newly minted pilots and specialists of all types, using the newly crafted airplanes as they slowly became available, for the coming battle over Europe.[5]

In 1942, after the group had been deployed to England, LeMay developed the aircraft formation used by the prewar Boeing B-17 bomber groups into that formation described and depicted below. Fundamental to the success of the U.S. strategic bombing on Germany was adoption of the policy for airplanes to discontinue all evasive maneuvering between the Initial Point (IP) and the target, until bombs were released, so the bombardier could line up the bomber on the target to achieve the best possible result. LeMay convinced his crews that the evasive maneuvering that had been used at the time did little to affect their overall safety, but did much to lessen the accuracy of their bombs.[6] Following steady improvements by the 305th utilizing these changes to procedures, all bomb groups changed to these tactics.

With no combat bombing experience, the relatively new U.S. Army Air Service, later the Air Corps, and still later the Army Air Force, had to develop all command structure, supporting units, and the tactical procedures necessary to allow routine application of this new type of military force as the senior leaders might direct.

It might seem easy to fly over something thought of as large, like a factory or an airfield, and drop a bomb on it. But to do so, when and where needed, to cause destruction of that target, and to do so with the smallest overall risk to one's own forces (people and equipment), requires planning and organization. This planning and organization that go into a single bombing operation allow these operations to be repeated with the expectation of similar results for each operation. In the period of the 1930s and early 1940s, the U.S. Army Air Corps established the process and procedures described below.

World War II Example

The command structure of the air forces which had the responsibility for conducting aerial bombardment against German-held areas of Europe during World War II was the 8th Air Force. The bomber forces were contained in three air divisions, in turn consisting of multiple combat wings, each wing containing three groups, and each group normally consisting of three squadrons. A squadron nominally had twelve aircraft, of which something like half would be ready for a single mission.

The group was the basic operational unit at that time. The higher-level organizations each had a staff to perform preparation tasks common for all the units under it. The staff at the numbered Air Force level compiled weather reports, maintained the target files and attack plans, and determined the quantity of bombs and best fusing for each target. There in the Operational Research section, a staff of specialists studied the aerial photos and other intelligence available to determine the best application of bombs to destroy each target complex. These specialists determined the proper fusing of the bombs to cause maximum destruction of the selected target. In the Operational Intelligence section, which maintained the target files, the Aiming Point for a target was identified. This point was normally a distinctive building near the center of the complex.

The actual targets and their relative priorities were determined by even higher-level staff agencies. For World War II Germany, a bombing priorities committee in London had established the priorities for all industrial and military targets in Germany. The bomber commander would generally assign the highest priority remaining target from the list, for which the forecast weather was suitable, and the bomb load requirements achievable with the anticipated number of available aircraft. The overall plan, including primary, secondary, and last-resort targets, size of the attacking force, general route of flight, and time of the attack, was communicated to the Air Divisions.

The staff at each air division took the overall plan and fleshed it out with assignment of units, adjusting the general routing specifically to better avoid enemy defenses, assigning navigation points as recommended by staff officers, and determining the time of assembly for the formations. The division staff was aided in their decisions by large wall charts tabulating all the assets of the division with tables showing status of aircraft and crews. These status boards were kept continuously up to date so that all decisions were based on current information.

The combat wing staff took the plan as communicated by the division, and worked out the takeoffs and formation assembly portions of the flight, and the specific bomb load for each type of aircraft participating. This staff's job was to get the various groups into the air and to get them assembled into the larger formation. To do this, they assigned particular groups to provide the aircraft and crews for specific positions within the formation. The order from division had specified the fuel load, and the wing determined the number of bombs that each airplane could then carry.

In the offices of each bomb group involved in the mission, the staff officers responsible for intelligence, operations, weather, communications, and navigation, along with the staff bombardier, began preparations for passing all the detail information communicated from the higher-level organizations to the air crews who would perform the mission. Packages containing maps, photographs, and weather charts and data were assembled to be provided to each individual crew. The crews all assembled in a large briefing room for a mass brief of common information for all. The initial portion of this briefing covered the primary target, the route, and the related times. The Initial Point (IP), the runup course to the target, the Aim Point (AP) of the target, and the Rally Point (RP) for the mission, were identified and depicted on a projected screen. Also illustrated with the help of a projected map was the target's position in relation to visible landmarks such as built-up city areas, rivers, rail lines, and highways.

Operations and intelligence officers continued the briefing by specifying the bombing altitude and describing the target in detail. Such a mission would expose thousands of crew members to the dangers of the enemy's defenses in addition to the basic risks of a large-scale aerial operation over long distances, making it essential to properly motivate the crews. As a result, the briefing included explanation of the value of the target to place it in context of the larger conflict under way. For example, the destruction of a combat aircraft manufacturing plant would tend to reduce the enemy's ability to operate those aircraft in the future. A similar argument would be made for a ship-building yard, or any munitions plants. Additionally, other operations being conducted in support of the bombardment mission, such as attacks on the enemy's air defenses, placement of search and rescue resources, and emergency procedures were also included both to inform the crews of these preparations and to demonstrate the importance of the mission.

The weather officer continued the mass briefing with that topic. The general weather situation was illustrated on a large wall map of Northern Europe. The forecast covered England, France, Germany, Poland, Denmark, and the North and Baltic Seas. Frontal regions were also illustrated on the map. A chart was projected showing forecast cloud height, cloud cover, winds, temperature, visibility, and icing conditions along the route of flight from the departure base to the target. The chart also showed from the surface to 30,000 feet, in altitude increments of 5,000 feet. This forecast was discussed spanning the time from departure to landing back at base. Of note was the forecast temperature at the bombing altitude, as well as the freezing level on the route out and for the return leg.

Next, the flak officer illustrated the known enemy antiaircraft gun installations along and near the route of the day's mission. This was followed by an operations officer briefing the bomb load and fuse settings for the bombardiers. He continued to the formation assembly portion of the flight. The flak briefing covered the assembly altitude and course, and the course to the wing assembly area. Also, this section included illustration of supporting elements, where other friendly aircraft would be operating that day, and the locations where escorting fighters would be near the bomber formation. Specifically illustrated was the IP location, where the group maneuvered into the formation for the bomb run.

Following these briefings of common topics for all crew members, the individual crew position types gathered with the group staff officer of each specialty. Pilots, navigators, bombardiers, radio operators, and gunners met for highly technical and specialized briefings on these specialties. These sessions assured that all crew members had complete preparation for the day's mission.

Following the briefings for all crews, the crew members donned their flying clothes, particularly those items designed to protect them from the terrific cold they would experience in the flight at altitude in unpressurized airplanes. After storing their personal items and receiving escape materials, they were transported to the aircraft parking areas. Here the final briefing occurred between the pilot/aircraft commander and the specific crew for that particular airplane.

Engines were started at a time announced in an earlier briefing; the first airplane taxied toward the runway at a specified time, and a flare signal from the control tower informed the lead airplane to begin the takeoff. Each airplane sequentially took the runway to become airborne.

Following takeoff, the lead plane in each element flew straight ahead for ninety seconds, then made a distinctive bank to the left, to mark the turn point, before settling into a gentle left turn. The following planes of the element turned when the plane ahead turned. The three-plane element continued the left turn and proceeded away from the field opposite to the direction of takeoff, while climbing to 1000 feet. Another left turn was made to bring the element, now formed into the inverted "V," often called a Vic, with 50-foot separation in altitude at two miles and parallel to the runway. At this stage, two elements joined to complete the squadron. The relative position of the second elements depended upon where in the group formation the squadron would be; if in the lead squadron, the second element leader flew with the nose just behind and to the right of the rightmost aircraft of the lead element. If in either the high or low squadrons, the second element leader flew just behind and to the left of the leftmost aircraft of the lead element.

As they continued to circle the field, they climbed to 2000 feet, where three such squadrons joined in the group formation. The lead plane of the joining high and low

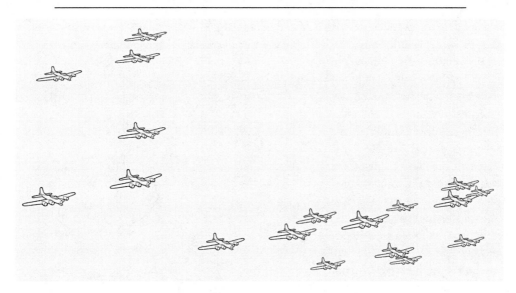

Perspective view of the group formation seen from above and to the side.

squadrons moved in to fly even with the lead plane of the second element of the lead squadron.

Vertically, the elements of the lead squadron were stacked with the leftmost plane highest and the rightmost plane lowest. The elements of the high and low squadrons were stacked with the rightmost plane highest and the leftmost plane lowest. The second elements were always lower than the first elements.

When the formation was complete, the group proceeded to the wing assembly point. The wing was assembled similarly with a lead group, and a high and a low group. The lead aircraft of the low group moved in to become aligned with the tail and to the left of the rearmost aircraft of the lead element of the lead squadron of the lead group. Similarly, the lead aircraft of the high group moved in to become aligned with the tail and to the left of the rearmost aircraft of the lead element of the lead squadron of the lead group. The assembled three groups of the combat wing fill 3000 feet of altitude. The wing then proceeded to the rendezvous point to join with any other wing assigned to make up a larger task force. Individual combat wings would form in trail with about 60 miles of separation.

The combat wing of three groups had been determined through many trials and exercises as the optimum scheme for the formation to defend itself with the guns mounted on the aircraft. This pattern provided for the maximum useful field of fire for all the guns of the formation.

The assembled combat wing proceeded along the briefed route to the initial point or IP. The IP had been chosen as a turning point to place the airplanes onto the bomb run. That is, it is the end of the en route segment to the target area and the beginning of the bomb run segment. The staff planners selected the IP relative to the target so that the course between IP and target, called the runup, would enable the bombardier to positively identify the correct target. Following the turn onto the runup, initially the groups of the combat wing maneuvered into the formation used during the actual bomb delivery, and were then committed to this course until after the bombs had been released. This formation did not provide as much defensive advantage as the combat wing, so it was important that this leg was just long enough that the maneuver could be completed and the planes were steady on the new course, but not much longer because of the increased exposure to enemy defenses.

As the lead aircraft in the lead group reached the IP, it fired two red flares to signal the entire formation, then turned onto the runup course. This was one of the high points of the mission for the lead navigator and bombardier, as a failure to reach this point seriously jeopardized the entire mission. The low group remained on the en route heading for an additional twenty seconds before turning onto the runup course. This displaced the low group nearly a mile to the side of the lead group. The high group remained on the en route heading for an additional twenty seconds beyond the point where the low group turned before turning onto the runup course. This displaced the high group nearly two miles to the side of the lead group. While the groups were spread out in this manner they were at their most vulnerable time to the enemy defenses.

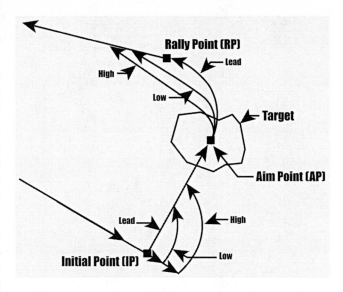

Diagram of the relationship of the Initial Point (IP), the Aim Point (AP), and the Rally Point (RP) for a bombing target.

All the foregoing efforts came down to the crucial next several minutes; the pilot passed control of the airplane to the bombardier. The bombardier now used the ground references prompted by the planners to assure that the plane was on the correct course. He used the Norden bombsight to locate the target aiming point ahead, applied adjustments through the instrument to account for affects to the aircraft's ground track and speed, and monitored the instrument to assure that the predicted track of the airplane passed precisely over the aim point. The bombs could be either released manually or automatically by the bombsight. This striving for precision in the release was because any error in the release point generally increases while the bombs fell through the air. With these methods and equipment, a good crew was expected to place bombs within 300 yards of the intended point. The variety of enemy actions to stop the attacks caused many bombs to land beyond this distance, being offset by a larger attacking force.

After the bombs were released and the bombardier had closed the bomb bay doors, the lead plane in the group continued straight ahead on the bomb run heading for fifteen seconds. He then began a gentle turn toward the Rally Point, the point where the group would reestablish in the same combat wing formation for the increased defense it provided for the return trip to home base. The radio operator in the lead plane transmitted a message to the Bomber Command communicating the time of attack on the target. The following two groups rejoined with the lead group as soon as possible, regularly cutting the corner between the target and the position of the departing lead group.

With the turn toward home base made, the bombers were now subject to the full fury of the enemy fighter force. The route planners did their best to make the homeward course free of antiaircraft artillery, but with the bomb run completed, there were no longer any unknowns about the intentions of the attacking force. Prior to the bombers reaching the IP, the enemy defenses were on alert waiting to learn the true target; i.e., to learn which of the fighter bases should engage the attacking bombers. During this period certain fighters were presumably tied up, pending the notification they would be required to defend a particular target. Once the attack was complete, specific targets no longer needed defense, and all fighter units could be assigned to attack the departing bomber force; it was up to the enemy defense command to determine how much fighter force to apply to the highly predictable return course of the bombers. Influencing these defensive decisions by the enemy was a powerful reason for coordinated attacks by multiple bomber formations in different places, at different times.

The final phase of a bombing mission was the interrogation of the combat crews after they arrived at their home base; it was back to the briefing rooms at the group for all uninjured crew members. Those who had something urgent to report, such as having observed friendly planes in trouble or going down or other crew members bailing out, did so immediately upon arrival. This information was passed immediately to Flying Control and on to search and rescue centers.

Each crew gathered as a crew around a table with a debriefing officer. This officer conducted the interrogation, often asking leading questions to help assure that all pertinent information was brought out. Crew members related observations and events relative to navigation points, altitude, and time during the flight. The bombardier provided an assessment of the quality of the bombs on the target. All encounters with enemy defense elements were discussed in great detail; the interrogator was interested in the number, type, method of attack, and markings on the fighters, which were all detailed. Gunners reported their results in enemy planes shot down or damaged. The gunners made any specific claims of enemy aircraft destroyed or damaged. The debriefer questioned the crew about antiaircraft artillery: their location, numbers, accuracy, and any new methods observed. The crew was prompted to describe any military installation or construction observed along the flight path either going to or from the target. Collecting these reports could be invaluable in the preparation of another attack in the future.

When all the crews had been similarly debriefed, the group staff completed the interrogation forms and compiled the basic statistics of the group's part of the day's mission. The staff assembled and generated the group report on the bombing mission, which was sent to the next higher headquarters. The communication flows used in the planning phase of the operation to deliver the plan from command down to the bomb groups, now were used in reverse; results and observations obtained from the crews were sent up the chain for evaluation. At division staffs, the reports were the focus of critiques. Of specific interest in these critiques was any flaw during the mission and the planning to prevent such failure on future operations. The reports received from the people who flew the mission were used to update the planning charts at all levels to indicate the latest and best understandings of enemy defense unit locations and capabilities.

Strike photography captured following bomb release was developed at the group and reviewed by trained photo interpreters as soon as possible. Although results could be obscured by smoke and dust, these photos still provided both commanders and planners with an immediate assessment of how well the attack covered the target and the extent of the damage. A separate photo mission flown after the smoke and dust had cleared, the following day, would provide evidence of the actual degree of destruction achieved. These photos were used to plan any future mission on these specific targets, and became part of the target file so that intelligence staff could monitor activity at this target following the strike.[7]

The flight of a bomb group from England to a strategic target in Germany illustrates the capabilities of the B-17 force. Total distance from the departure base to the target area is a little over 600 miles. The set of navigation points to get from eastern England to a target complex on the outskirts of Berlin were selected by the planners as distinctive and recognizable as viewed from the 25,000-foot en route altitude the planes would fly. The intersection of two major highways, the intersection of a river and a canal, a railroad intersection or marshaling yard, or a prominent transmitter site are some of the most common points chosen for the pilotage task.

One might ask the question: How would American crews know such landmarks in Germany? The answer is, on the first couple of trips they would not. They would have gotten some insight from the RAF, who had been flying over Germany for over two years at that point. And they simply would have started to accumulate the knowledge. After flying over an area, you become familiar with it; the town is at the bend in the river, the seaport is in the protection of the harbor, rivers require bridges. Human civilization has established these relationships the world over, primarily for the ease and comfort of the population before modern transportation tended to make them less of a requirement. Cataloging of such prominent navigation points was one of tasks of the debriefing officers derived from the comments of the returning crews.

The first two hours after departure were consumed with the squadron, group, and then combat wing formation assembly maneuvers. During this time, aircraft with mechanical issues moved out of the formation and returned to base to be replaced by spares that were included in the mass takeoff. Other aircraft of the planned formation might have returned as well. For example, if a group was unable to rendezvous properly with other groups, it would return. The timing of the various joinups was critical, and if a group could not be in position at the right time, it may never catch up with the larger formation as it moved toward enemy territory. Planes not in the planned formation were quite vulnerable; additionally, they would not contribute to the common defense. After it was determined that the remaining ships were ready, unused spares returned as well.

Having assembled into combat wings over England, the bomber force would depart the English coast for a landfall on the continent of Europe. For example, to Berlin, this landfall would be on the Dutch coast, near Egmond. This would be followed by one or more segments to reach the IP as described above. The number and direction of these segments would vary as they constituted efforts to reduce exposure to enemy defenses. Additionally, the direction of these segments was intended to keep from the defenders the location of the ultimate target for as long as possible. For the Berlin example, these en route segments would consume just over three hours of flight. The time spent in the target area dropping bombs, the reason for the trip, lasted less than twenty minutes in this case. The return trip would take a similar three hours' time. With the departure and assembly, en route, IP to target, return, and final descent segments, the total flight time for this example is over eight and a half hours.[8] This was not the longest mission possible for the B-17, but nearly so. A longer mission by the bombers resulted in less fighter escort coverage in the target area.

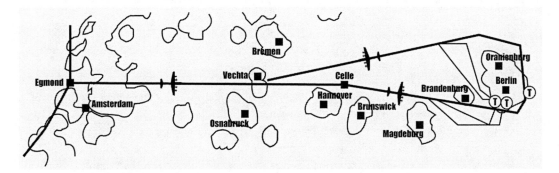

Depiction of the flight route across Europe to targets in the Berlin area.

Each B-17 bomber described above carried ten 500-lb. bombs. This meant that each aircraft carried 5000 lbs., each group delivered 90,000 lbs., and each wing 270,000 lbs. A single target, such as an aircraft factory, would be attacked with commonly three such bomb wings or 810,000 lbs. of high-explosive bombs. In terms of tonnage, this is 405 tons. For comparison with later sections in this book, this is 0.4 kiloton (KT).

The application of these premises to the bombing campaign against Germany, and the German-occupied lands at the time, initially proved less than a complete success. Several iterations of improvements were made to the bomber formations to maximize the protection provided by the guns, improve the concentration of the delivered bombs, and move more and more aircraft to the desired target.

A significant lesson for the Americans was the need to defeat the German fighter defenses to allow the bombers unimpeded access; the defensive guns of the bomber formations were inadequate. The American and British escort fighters available at the time were unable to provide protection over the deeper areas of Germany from their bases in England through 1943. This changed with the availability of the longer-range North American P-51 Mustang, used by both America and Britain in this fight. By the time that these longer-range escorts were available in quantity, the pipeline of bombers and crews was also adequate, enabling a sustained attack on the German war-making industries.

Being recognized for these innovations that became fundamental to the strategic air campaign in Europe, LeMay was moved to the Pacific as commander of all air divisions in the bombing operations of that theater. In this assignment he was to refine existing groups and incorporate new ones flying the brand-new and initially troublesome Boeing B-29 Superfortress aircraft. Shortly after his taking over, the U.S. advance across the Pacific had yielded bases in the Marianas Islands group: Saipan, Guam, and Tinian. While airfields were being prepared on these islands, Iwo Jima, about halfway between the Marianas and the Japanese home islands, was taken as well. This outpost provided a base for supporting escort fighters as well as an emergency landing site for B-29s needing refuge while on the 3000-mile round trip from the Marianas to Japan.

This buildup of bombing capability resulted in most probably the greatest testament to the concept of strategic bombing: Japan was induced to surrender without a ground force invasion of Japan proper. The air campaign culminating with the dropping of two atomic weapons laid such waste to Japanese cities that the government capitulated. Although this city destruction was not universally popular with American purists who held that strategic attack should be directed at war-making industries and infrastructure, it certainly met the definition provided by Douhet.

Norden Bombsight

The combination of optical sighting, gyro stabilization, and mechanical computing was the key to an effective bomb aiming system.

The bombing mission was beset by conflicting needs: the need for accuracy to destroy a target, and the need to protect the bomber and crew from exposure to the enemy. The accuracy objective was best accomplished by a low-altitude attack. The avoidance objective was best accomplished by higher altitude, i.e., remain above the altitude reachable by the enemy's defensive weapons. The potential errors in the delivery of aerial bombs were largely angular, chiefly in the sighting angle from the approaching aircraft to a target. Because any error in this angle resulted in a larger horizontal miss distance on the ground as altitude increased, greater accuracy was favored by a lower altitude. Because the bomb, once released

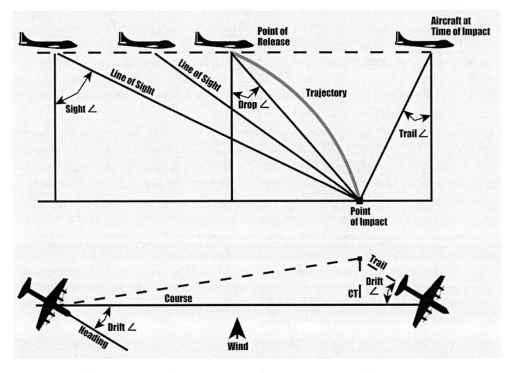

Diagram of the bombing problem illustrating the sighting angles, the trajectory of the bombs, drift, and the trail.[9]

from the airplane, was slowed down by air resistance as it fell, it landed somewhat behind the aircraft, which had continued onward at a near constant speed. This created the requirement to identify the point in advance of the target where release would land the bomb on the target.

To minimize these errors with altitude and to standardize the procedure for bombing, the mechanical/optical gunsight was adopted. The primary instrument used by U.S. Army Air Forces in World War II was the Norden bombsight, named for Carl Norden, who developed the design and founded the company which was to provide this equipment to the Army during the war.

The sight could look forward of the aircraft 70 degrees. This meant that the sight could be placed on a target several miles in front of the aircraft, depending on altitude. For example, for a B-17 to bomb from 18,000 feet, the target could be set up at a distance of 8 nautical miles, and the resulting bomb run was 3.75 minutes at an aircraft speed of 150 miles per hour.

A vertical and a directional gyroscope provided a stable reference for the bombsight's calculation of angles, which were referenced to the vertical axis of the directional gyro. The principal of a gyroscope is that a spinning disk resists forces that would perturb the position of the spin axis. As a result, short-term influences, such as the pitching and rolling of the airplane due to both pilot inputs and the turbulence of the air, do not affect the axis. So the axis position is maintained in the established reference position. This instrument established the vertical direction reference for the angular calculations.

The mechanical computer consisted of a variable speed drive that could be set up to advance gear mechanisms at the rate the airplane would move along the bombing course at given altitude, airspeed, and wind conditions. The mechanisms moved at the correct rate included a mirror that deflected the optical path of the sighting telescope to maintain the sight crosshairs on the target, and angle measurement index pointers. The sight angle pointer moved closer and closer to the drop angle pointer as the plane moved along the bombing course. Eventually the two pointers touched, and a signal was sent to the bomb racks for release of the weapons.

Information was tabulated for the reasonable range of speeds, altitudes, and weapons types for the bomber aircraft of the day. Information in these bombing tables was: DS or Disk Speed, ATF or Actual Time of Fall, Trail, Drop Angle, and Sight Angle.[10]

The disk speed is a number needed by the internal bombsight mechanism that accounts for ATF, and hence a combination of aircraft ground speed and Bombing Altitude. ATF is the time it takes that type of bomb to fall from the selected altitude, starting at the selected airspeed, to 5000 feet. Trail is the distance, expressed in mils or milli-radians, that bombs would land behind the aircraft. The values of Drop Angle and Sight Angle are the angles that will be indicated by the pointers on the bombsight when this value of ATF and Trail are set into the instrument. The tables are entered with the bomb to be used, the Bombing Altitude, and the True Airspeed.

The numbers in the bombing tables are for the case where the target is at 5000 feet elevation, and there is zero wind, hence there is no drift. The tables contain values for adjustment for each 5000 feet of elevation difference, and for each 10 degrees of headwind or wind drift. With these adjustment factors, the best estimates for target elevation and wind are used to calculate interpolated adjustment values. The table values for Disk

Speed, Drop Angle, and Trail are adjusted for actual target altitude and winds before bombsight setup.

Picture of a Norden Bombsight showing the eyepiece and the stabilization element (author's collection).

Elements of the bombsight operation, and the inputs to the bombsight were:

Stabilization	— Leveling Knobs
Aircraft Course	— Turn Knob
Sighting Angle	— Search Knob, Rate Knob, Displacement Knob
Disk Speed	— Disk Speed Knob
Trail	— Trail Arm
Wind Drift	— Drift Knob

It was the task of the bombardier to plan the bombing run, calculate the required adjustments from values in the bombing tables, input the values into the bombsight controls, make the final manual adjustments for incorrect wind speed and ground speed inputs, and monitor the bombsight to assure that the selected target was correctly tracked by the optics.

The first step in the setup of the bombsight was the leveling of the gyroscopic reference. The bombsight includes two spirit levels and adjustment knobs to enable the operator to observe the state of the levels at any time and make adjustments if needed. Setting the adjusted values for Disk Speed and Trail provide the computer with the values to represent aircraft altitude, airspeed, and bomb characteristics. The horizontal crosshair is set on the target with the Search Knob to establish the target range. This range setting is fine-tuned with the Rate and Displacement Knobs. The vertical crosshair of the optical sight is set on the target with the Turn Knob to establish the aircraft course. If there is crosswind, the Turn and Drift Knobs are both moved in a step called killing drift. This step adjusts both aircraft course and pointing of the sight, establishing a drift angle that can be read on the Drift Scale. If the crosshairs do not remain fixed on the target, indicating that the prediction of the motion by the bombsight computer is not perfect, the bombardier uses the Displacement Knob to adjust the horizontal crosshair, and uses the Drift Knob to adjust the vertical crosshair. These adjustments have the effect of altering the ground speed and/or wind speed and direction values in the total representation of the aircraft's speed over the selected range to the target.

The correct setup was indicated by the crosshairs of the optical sight remaining fixed on the selected target. This was a uniqueness to the Norden design and operation. For other bombsights of the era, the optical crosshairs remained in a fixed position and the bomb aimer observed the target move under the crosshairs when over the proper release position.

The bombsight continuously determined two angles: the range angle and the current angle to the target, which were displayed to the operator by the Sight Angle and Drop Angle on the index pointers. From these it determined the angular movement needed for the optics to remain on the selected target. When the difference between the two angles became zero, the bombs were automatically released. The bombsight generated outputs that controlled a panel meter visible to the pilot indicating that the plane was left, right, or on course. Similar signals were routed to the autopilot for automatic control of the airplane's course on the bomb run. Most often, during the final portion of the bomb run, the pilot handed the airplane over to the bombardier so that the final minor course corrections were made by the autopilot under the close monitoring of the bombardier.[11]

Building SAC's Global Bombardment Force

As with the previously described buildup of defenses to protect from air and submarine attacks, in 1948 the U.S. began to once again expand its skeleton aerial bombardment military force. As with those other defensive forces, this action was in response to perceived intentions of the Soviet Union.

At the time of this renewed focus on aerial bombardment, Lt. General LeMay was in a postwar assignment as commander of United States Air Forces Europe (USAFE). While he was in this post, the Berlin Blockade and subsequent airlift of supplies began. After these operations had been initiated, it was decided that LeMay's talents would be ideal for the great emphasis that needed to be placed on the organization and training of long-range bomber forces. As a result, LeMay was assigned to command the approximately one-year-old Strategic Air Command (SAC). LeMay essentially had to repeat some of the organization development he had done in the previous wartime assignments, when the job was to take that bombing force that had survived the rapid and significant postwar demobilization and grow and train it into an effective Strategic Air Command. Now the U.S. was to have a standing combat-ready air force in other than wartime. This was a significant departure for the U.S. from the longstanding approach of maintaining only a skeleton force of professional military in time of peace, to be expanded by volunteers to create an expeditionary force when the situation required.

Two of the practices that LeMay had instituted as far back as World War II to improve bombing results, as well as overall operations, were the study of targets and the use of checklists. With the emphasis on radar as the aid to navigation at the end of and following the war, this included study and recognition by the navigators, of the radar picture received from the planned target. SAC became renowned for the use and adherence to the checklist, as the means of standardizing crew performance.

Groups were built up and brought to full combat readiness one group at a time.[12] Initially the force was equipped with remaining B-29s. These were replaced by B-50s, a more powerful improved design on the B-29 foundation. The first truly intercontinental bomber came along as the Consolidated B-36. Just as the B-29 had been initiated as a longer-range follow-on to the B-17 with greater payload and crew comfort, the B-36 had been initiated during the war to provide even longer range. This need for extreme range was established when it was believed that England could fall to Nazi attack, and the much-needed air bases located there would not be available for the conduct of attacks on Germany. The first all-jet bomber came in 1951 as the Boeing B-47. This bomber was acquired in quantity, eventually a total of 1700. The B-47 was followed by the faster, higher-altitude, higher bomb load, and longer-range Boeing B-52.

During World War II, as the ground-based gun defenses achieved greater range, there was the resulting desire for the bombing aircraft to fly at higher altitudes to avoid the defenses. Additionally, higher speed was desired to subject the bomber to the defenses for a shorter period of time. This trend continued after the war with the advent of jet aircraft; higher speeds, plus higher altitudes to avoid first the gun defenses and later the surface-to-air missile (SAM) defenses. By the 1960s, planned bomber aircraft were traveling at higher than the speed of sound at up to 70,000 feet altitude. The relatively new SAM defenses and the newer fighter interceptor aircraft could reach bombers at these altitudes and speeds. The lesson was that if the defenses could locate the bomber aircraft on radar, a weapon could be sent. This led to the sharp increase in low-altitude flying by attacking bomber aircraft, to fly for as long as possible "under" the view of the defender's radar.

An innovation that came out of the SAC buildup was the emphasis on inspection of units, focusing on their ability to execute their portion of a war plan. A significant difference between SAC and the somewhat equivalent bomber forces of World War II is that the World War II forces were conducting ongoing operations where actual results were being tallied and analyzed. This is a key outcome of the post-flight interrogation of crews. These results could be reviewed and increased emphasis, change to tactics or procedures, or training could be applied to bring about improvement. In the SAC Cold War environment, results of exercises had to be substituted for actual combat results. In order to evaluate a unit's complete set of contributing functions, standards for performance had to be established for virtually every task performed on a SAC base. Units practiced short notice takeoffs onto their war-plan flight plan routes, but it was the Operational Readiness Inspection (ORI) conducted by the highest level of SAC that measured the unit's capability and provided a score for comparison command-wide. A unit with low scores that could not be accounted for would expect a change in commander. Soon.[13]

The nomenclature of the basic unit had been changed from group to "wing" during this time, but this unit still consisted of nominally three squadrons of aircraft.

As the range of the bombers increased to the point of launching from the northern U.S. to targets in the Soviet Union, the need to provide additional fuel along the way came to the fore. This brought about creating a method and then the equipment to transfer fuel from one airplane to another in flight.

Air Refueling Tanker Evolution

Although aerial refueling for the U.S. military traces its origins to 1923, the ready availability of overseas bases during World War II deferred implementation until the late 1940s. With the advent of SAC's global mission, and the possibility that overseas bases could become unavailable, the U.S. put in place an air-to-air refueling capability for the strategic bomber aircraft. Over time other military aircraft, particularly jet-powered tactical fighters, were equipped with the capability to receive fuel from the tankers as well.

The U.S. experimented with three arrangements for the transfer of fuel: the grappled-line-looped-hose approach, the probe and drogue approach, and the boom and receptacle approach. The grappled line refers to a cable trailed from a receiver aircraft. The tanker aircraft fires a line to "grapple" this cable and pull it over to the tanker aircraft. This cable is attached to the refueling hose, which the receiver aircraft then pulls over and attaches it to its own fuel system. The probe and drogue trails a hose from the tanker aircraft fitted with a conical basket at the end of the hose. The receiver aircraft is fitted with a probe that it flies into the basket to connect to the tanker's hose. For the boom and receptacle configuration, the tanker aircraft is fitted with a trailing telescopic boom with small flying surfaces. The receiver aircraft is fitted with a fixed receptacle connection. The receiver aircraft flies into position slightly behind and below the tanker aircraft. With the receiving aircraft in position, the boom operator aboard the tanker flies the boom into position to connect to the receiver's receptacle. This is most often called the flying boom method.[14] All of these approaches have been used by the U.S. military; the latter two approaches are both used today. In fact, the U.S. Navy uses the probe and drogue method, whereas the Air Force selected the flying boom approach because it concluded it is better for transferring large quantities of fuel.

The initial aircraft chosen for use as a refueling tanker was the Boeing B-29. In 1948 a small quantity of B-29s were converted to KB-29 by the addition of the loop hose equipment for the grappled-line approach. This was the method used for a round-the-world nonstop flight of a B-50 in 1949.[15] When SAC selected the flying boom method, over 100 KB-29s were converted by installing the boom equipment. There were eventually over 280 KB-29s converted to be flying boom tankers. Later, B-50s were also made into tanker versions with the drogue equipment to refuel tactical aircraft equipped with probe refueling.

With the growing need, the Air Force chose the Boeing KC-97 Stratofreighter as the first purpose-built refueling tanker, which was equipped with the flying boom. This resulted in the highest quantity of dedicated tanker aircraft purchased at 814 aircraft total. The KC-97, the "K" for air refueling tanker, was adapted from the C-97 airplane. The C-97 is itself closely related to the Stratocruiser passenger airplane, which itself had some lineage to the B-29. Beginning in 1950, these aircraft served for many years, but much of that time in Air Reserve and Air National Guard units. The KC-97 was required to fly at maximum speed of 230 MPH in order to refuel the jet-powered B-52, which had to perform the operation at bare minimum speed. If taking a large load of fuel from the tanker, the B-52 was at a dangerously low speed for that weight condition, making the operation more dangerous than desired.

This danger was removed when Boeing developed a jet-powered tanker, which the Air Force bought in quantity beginning in 1957. The KC-135 Stratotanker was the result. The USAF eventually bought 749 of these four-engine jet aircraft, many still serving today.

The following aircraft provided the manned bomber and aerial refueling tanker capability during the early and mid periods of the Cold War. These aircraft would be replaced by later developments and would be retired from front line service by 1965.

Aircraft	Range	Speed	Weapons
B-29	2820 nm	252 knots	
B-50	2080 nm	212 knots	
B-36	3465 nm	200 knots	
B-47	1749 nm	484 knots	Two 3.8 MT, or one 9 MT
KC-97	2000 nm	200 knots	

Table of early Cold War bomber and tanker aircraft capabilities.

Boeing B-29

Powered by 4 Wright R-3350 radial engines, to cruise at 190 knots, at up to 32,000 feet, to carry up to 20,000 lbs. of bombs.

Range: 2820 nm

Boeing B-50

Powered by 4 Wright R-4360 radial engines, to cruise at 212 knots, at up to 37,000 feet, to carry up to 28,000 lbs. of bombs.

Range: 2080 nm

Consolidated B-36

This huge six-piston engine—propeller design was initiated during World War II when it appeared that England could fall to a Nazi invasion and the U.S. might have to be capable of attacking Germany from bases in North America.

Powered by 6 Wright R-4360 radial engines, plus 2 General Electric J47 turbojets, to cruise at 200 knots, at up to 43,600 feet, to carry up to 72,000 lbs. of bombs.

Range: 3465 nm

Boeing B-47

This six-engine swept-wing design had a crew of three. Virtually all large jet aircraft, both civilian and military since the 1950s, trace their heritage to the B-47. B-47s were rotationally based in England and North Africa so as to be in better range of targets in the Soviet Union.

Powered by 6 General Electric J47 turbojets, to cruise at 484 knots, at up to 33,100 feet, to carry up to 25,000 lbs. of bombs.

Range: 1749 nm

Boeing KC-97

The KC-97 was a version of the Stratofreighter transport aircraft, itself a derivative of the B-29 bomber. Fuel tanks were separated so that the piston engines would draw aviation gasoline and separate tanks provided for offloading jet fuel to the B-47 and B-52 receivers through the flying boom refueling method.

Powered by 4 Wright R-4360 radial engines, plus 2 General Electric J47 turbojets, to cruise at 200 knots, at up to 30,000 feet, to carry up to 9,000 gallons of jet fuel.

Range: 2000 nm

The following map depicts the deployment of U.S. strategic bombing forces in the spring of 1952.[16]

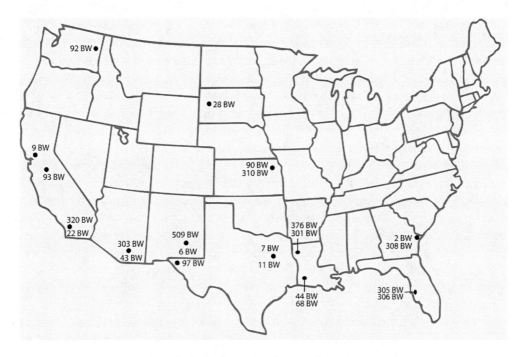

Map of U.S. strategic bomber unit bases in 1952.

With the B-29, B-50, and B-47 as the attacking bomber aircraft, reaching targets in the Soviet Union required either refueling aircraft based in other countries, or the bombers to be based in other countries, or both. With the B-36 and B-52 aircraft, their range capability lessened this requirement somewhat, but the basic operational scenario for the U.S. manned bomber attacking targets in the Soviet Union was for one or more air-to-air refueling events for each bomber.

In 1948 when the buildup of SAC was initiated, it consisted of 14 bomb wings, none at full strength, and only two flying the B-50 aircraft. A handful of B-36s existed, but the bulk of equipment by far was the B-29. By 1955 this had expanded to 23 wings operating the B-47, 6 wings operating B-36s, and one wing with the then brand-new B-52. Now including air refueling wings with KC-97s, the total came to about 2800 aircraft.

The mission of these units was primarily "training for global strategic bombardment." They were not routinely prepared to launch a nuclear attack immediately. For example, as a result of increased world tensions around the Hungarian uprising in 1956, some SAC bomber units were forward deployed with nuclear weapons. This resulted in the bomb wing located at Biggs AFB, Texas, to fly their B-36s to the base outside Albuquerque, New Mexico. There they obtained their bombs, which were stored at the Manzano storage complex near the Kirtland base. Following this, they departed for their overseas deployment base.[17]

BECOMING ALERT

What had been in place for the 8th Air Force in World War II England was a combination of telephone switchboards and teletype networks to communicate between command levels and dispersed sites. As described in a previous section, the planning for and initiation of a bombing mission had occurred over essentially a twenty-four-hour period. The field order received by the Air Division headquarters, from the numbered air force, had adjustments and additional detail appended prior to its being sent on to the Combat Wing. Similarly, the staff at the Combat Wing added the results of its planning before being sent to the Bomb Groups for execution. Telephone lines were used both in individual connections as well as conference setups to provide for discussion, clarification, and questions before and after transmission of the teletype field order.

This was adequate for missions fully planned only after the commanding general ordered a set of targets, based on weather, previous results, and new directives. The problem that the new SAC was facing was that with the threat of a surprise ICBM attack, the bombers could potentially be destroyed on the ground while waiting for the World War II–style mission planning cycle to complete. The response to this was threefold: pre-planned targets assigned to specific crews; planes and crews prepared and ready for takeoff in minutes; and essentially instantaneous communications to order launch of the aircraft from their bases when an attack on America was detected. These steps were embodied in the alert system initiated in 1956.

The policy at SAC had evolved to that of counterforce; that is, SAC was to wreak certain destruction on the enemy's military capability if the enemy launched an attack on any of the free world nations. This force was planned to provide a favorable outcome for the U.S. in any conceivable circumstance.

The U.S. adopted the ground alert scheme for the manned bombers in 1957.[18] This placed one-third of the nuclear delivery aircraft, and their supporting air refueling tankers, on alert for immediate launch at any time. The element of specific crews being assigned to specific targets, and their aircraft being fully loaded and ready to launch in minutes, came into being at the Alert Facilities located on SAC bases throughout the Western world, and in the missile launch facilities as they came into being. The number of Alert Facilities housing air crews grew to over one hundred. Additionally, the strategic bomber and tanker aircraft were dispersed so that no more than one wing of launching aircraft were pending on a single runway.

At SAC bases to the end of the Cold War, integrated crews were on ground alert to man their fully loaded airplanes and depart for their specific targets within fifteen minutes from the order to launch.[19]

This integrated crew shared the schedule of being on alert one week out of three. The week spent on alert could be excruciatingly boring for the members, and the next two weeks were occupied by normal training duties. Each bomber crew was assigned a specific target in the Soviet Union, or later in China. They studied repeatedly all aspects of this target and the navigation route to reach it. Particularly the radar navigator had to be intimately familiar with the radar returns he would observe at the key points along the route, and from the target area. The relationship of the actual target aim point to a distinctive object observable on the radar was established by the navigator by studying the terrain and likely by studying radar display scope pictures obtained by intelligence aircraft missions.

Each target had one entry in the larger target list. This set of information contained such specifics as target description, aerial photographs, map references, and possibly radar scope photographs. These individual items could be updated by intelligence agencies as new information was obtained. When the information was updated, the crew would have to study the new information and possibly revise the specifics of their plan to attack.

Each bomber on alert was loaded with a standard load of weapons. For example, a B-47 was loaded either with two 3.8 megaton or one 9 megaton nuclear bomb. A B-52 was most often loaded with four one-megaton nuclear bombs. A refueling tanker was most often loaded with maximum fuel: some portion for its flight to a predesignated point to meet one or more bombers, and some portion to be offloaded to the bombers.

The airplanes on alert were parked on the alert ramp, the outer security fence of which included the alert facility. Here were located quarters for the individual crew members, a dining hall, and recreation rooms for reading, watching TV, and similar activities. This building also contained the command post for the unit, which would receive the launch notification from the SAC Command Post via the Primary Alerting System. The entire alert ramp was a high-security area, with controlled access and patrolling security forces.

Although the week on alert was seven days of twenty-four hours when the crew had to be together, or at least very close such as in the rooms of the alert facility, they could leave the facility as a group. For this they had a truck including a driver at their disposal. This allowed for them to visit some other areas of the base such as the base exchange, movie theater, or recreational facilities. In this way it was possible for them to see family, albeit staying nearby as a crew. At some bases, recreational facilities such as a softball field would be adjacent to the alert facility, although outside the security area. This allowed for alert crew members to potentially continue their normal activities during their alert assignment, and offered another distraction to others on alert. An area of the base movie theater was reserved for alert crews as something like a benefit. At least once during the alert period, they could expect an exercise intended to test their preparation and readiness.

At many bases, the alert ramp was in what was called a Christmas tree configuration. This was a set of short aircraft taxiways positioned at 45° from a central taxiway that led directly to the end of the main runway. This Christmas tree pattern often had three taxiway-parking spots on either side of the central taxiway. This arrangement placed the ready aircraft just hundreds of feet from the end of the runway when the time came.

When the alert klaxon sounded, the crews would run out of the alert facility either to their waiting aircraft or to their waiting truck, depending on the distance from the building to the aircraft at a particular base. For a crew elsewhere on the base when the klaxon sounded, they would rush to their vehicle and quickly drive to the ramp in priority traffic lanes set aside for their use.[20]

Each SAC unit, i.e., a base, had an assignment of how many planes of what type to have on alert. A base might have bombers, tankers, or both on the alert ramp. Three or six bombers were common alert forces for a particular base.

SAC COMMAND POST

As SAC established the capabilities to conduct long-range strategic attack on short notice to almost any part of the world, the need evolved to have a central command and control establishment to oversee this capability.

The pre-planned targeting and the instantaneous communication elements came to reality in the SAC Command Post, an element of SAC Headquarters at Offutt AFB, near Omaha, Nebraska. Omaha has often been called the center of the continental U.S. Also located within this command post was the Joint Strategic Target Planning Staff (JSTPS), the joint service agency responsible for identifying and prioritizing strategic targets which U.S. forces might be called upon to attack.

The SAC Command Post was a three-story underground structure, adjacent to other headquarters buildings at Offutt AFB. This building could be isolated from the outside world if an attack threatened. Contained in the underground spaces were equipment to provide fresh air through a filtration and decontamination system, fresh water from deep underground wells, electrical power generation should the commercial supply fail, and stocks of emergency food to sustain the staff for some time. The entire building was constructed of thick concrete roof and walls and buried underground to withstand attack.

Access to the facility was controlled by one entry point for all the underground levels and an additional entry point for the sacred lowest third floor, each manned by selected security personnel. For entry to the third level, an individual had to be personally recognized by either the entry point security person or someone else inside the facility.

Offices located on the top two floors that provided support to the SAC commander and the staff were:

a. A weather center constantly collecting information from around the world.

b. A teletype relay center that sent and received messages from all SAC units and other commands, including USAF headquarters.

c. A communications status center that monitored and maintained all the communications circuits for the facility.

In the most secure third floor below ground was contained the stations and people who actually provided command and control of SAC's forces. Here messages were received from units the world over containing status on the readiness of aircraft, missiles, and crews. This information was fed into computers where it was used to perform analyses of SAC's state of readiness versus that of an adversary. Much of the computer-stored data could be utilized to generate large displays for projection on the forward screens for briefing purposes, and so that the entire staff was informed. Also on this most secure level were located:

a. An intelligence center.

b. The previously mentioned JSTPS.

c. A staff of communications controllers who controlled a vast set of radio communications facilities throughout the world. When the Command Post needed to communicate to any SAC aircraft in the world, these controllers had the means to route to the correct transmitter, select the proper frequency, and orient the antenna to achieve the needed communication path.

d. A staff of materiel controllers who could provide immediate assistance and guidance on the operation of SAC's wide variety of equipment, both on board aircraft and located in any of the command posts. These specialists could contact equipment design engineers if that had become necessary.

The critical command and control of the SAC Alert Force resided at the consoles of the duty controllers and the senior controller. The lead duty controller, a major or lieutenant colonel, was supported by two noncommissioned officers. These controllers were selected for their broad experience in SAC's bomber, missile, and refueling operations and procedures. The controllers' consoles had telephone and radio communications to all units. Through these communications paths, the controllers received information regarding all SAC aircraft and missiles including airborne alert aircraft and those aircraft deployed to forward bases as part of "reflex" deployments. The senior controller was a full colonel authorized to perform the day-to-day control of the SAC forces on behalf of the SAC commander. To do this his console contained several communications panels. The gold phone connected this console to a network with the National Military Command Center, the major military commands, and the highest level government personnel including the president and Joint Chiefs of Staff. If the president issued orders to prepare for or go to war, those orders would have come over this phone. With the red phone, the senior controller could communicate immediately either individually or in conference with subordinate SAC commanders, headquarters USAF, and the North American Air Defense Command in Colorado Springs. The SAC commander was always within reach of the gold or red phone, or a radio extension, 24 hours a day regardless of where he was. On the gray phone, the senior controller could communicate with any individual SAC unit. For example, if an airborne SAC aircraft were to have an emergency, the senior controller received this call and acted as a single point of contact to provide assistance.

When the SAC forces were to be alerted, the primary example being in time of war, the Primary Alerting System was in place for that purpose. On a panel at the senior controller's station, the sequence was initiated by pressing the alert button, which lights an indicator on the panel showing an active circuit to each individual command post. This caused a warble tone to sound at all SAC unit command posts, indicating a message was forthcoming. Every unit's controller copied the message, checked it for validity, and if it pertained to that unit's force, initiated immediate action. For aircraft units, this meant activating the klaxon, which sent aircrews sprinting to their aircraft. For missile crews, this meant initiating the checklist sequence that would result in missile launch. At the completion of the message, each receiving command post acknowledged receipt and understanding by pressing the acknowledge button on their respective consoles, which caused the light on the senior controller's console to go out. If a light did not go out, that command post was contacted via the gray phone to assure that the message was received and understood.

The alert message was prepared by one of the duty controllers, and read by the opposite duty controller in the presence of the senior controller, over a special red phone at the senior controller's panel. Each message that was transmitted over the Primary Alerting System was copied to the same receiving units over teletype. The system transmitted a test signal on all circuits of the Primary Alerting System, every three seconds, to ensure that each circuit was always ready. If a circuit was faulty, a status light on the senior controller's panel was lit. Work to restore a faulty circuit would begin immediately.

Other equipment near the controllers' stations were television monitors for local weather conditions and to verify identity of people entering the third level, equipment displaying the warning information of the North American Air Defense Command's Ballistic Missile Early Warning System (BMEWS), and the display of the Bomb Alarm System. The BMEWS system consisted of three radars located in Alaska, Greenland, and England, which provided warning of missiles launched from the other side of the world. This system plotted the flight paths of detected missiles, and if a path were headed for an American target, would display the number detected, the time to next impact, and the predicted impact areas. The predicted impact areas would be shown on a map as ellipses, indicating some uncertainty in location. This display also showed the tracks of unidentified aircraft flying in the atmosphere near North America. The Bomb Alarm System showed the status of numerous sensors located throughout the U.S. in expected target areas; the sensors would be activated if a nuclear detonation were to occur. These warning displays are replicated on the command balcony, where the SAC commander and his senior battle staff convened for briefings and actual emergencies. These positions had access to all the communications capabilities of the command post, particularly the red and gold phones. From his desk, the SAC commander could also address the entire staff of the command post via a public address system.

As an example, if the BMEWS system provided a detection of missiles predicted to land in the continental U.S., the senior controller would pass this information to the SAC commander. The SAC commander would order the aircraft launched into "positive control." The duty controllers would prepare and send the order through the Primary Alerting System. At all alert facilities, the klaxon would sound, sending aircrews to their airplanes, where they would begin their checklists to start engines and taxi for takeoff. They would then take off toward pre-planned positive control points located far from areas of enemy radar coverage. The tankers would proceed to prearranged refueling locations. These aircraft would all be airborne and safe from destruction by missile attack within a few minutes. However, they would not yet have an order to proceed to and launch their weapons against any targets.

In this example scenario, the same communication networks described above would be used to notify the president and other leaders of the missile launch detection. All of these facilities would be constantly checking the status of the detected threat and the system equipment, to prevent action based on a false alarm. Just as American missiles would take something like thirty minutes to reach their targets, missiles targeted on the U.S. would take a similar time to impact. This gives the president time for some consultation with advisors if an order is to be issued prior to any impact.

If the president were to initiate the order to launch U.S. missiles and bombers, either airborne or on the ground, these same communications links would be used. The order from the president would pass to the Joints Chiefs of Staff through their command post in the Pentagon, to the commander in chief of SAC, and hence from the SAC Command Post via the Primary Alerting System to all units, as described above. If any of these officials believed the order might not be correct, they could question before forwarding the order. As a result, all the officials in this chain of command had to concur with the order, or at the very least not object to it sufficiently, for the order to be transmitted to the crew members who would ultimately arm and initiate final delivery of the nation's nuclear weapons. For aircraft already airborne, this order would come through SAC's widespread set of radio transmitters. The final check would be done by the crewmen on individual bomber aircraft and in underground missile launch control facilities. Pairs of crewmen would use information entrusted to them separately to check that the orders were valid. For both bombers and missiles, these two-man teams could only then arm and launch the weapons.

This central command and control was fundamental to the nation's readiness to initiate a response to any perceived threat. But like the ready-to-launch weapons themselves, in order to provide deterrence, it had to be available continuously. This required a backup plan should some of these facilities not be available for any reason. Backup consisted of a succession plan for the command post of one of the subordinate commands of SAC to take over, should the SAC Command Post at Offutt AFB ever be out of action. Beginning in 1961, this backup was extended in the form of an Airborne Command Post which was airborne in a position where it could conduct all required communications 24 hours a day, 7 days a week. The Airborne Command Post carried a full staff of duty controllers and communication gear to perform the identical duties as described above for the underground staff. These controllers did duty at both the underground command post and the airborne version, on a rotational scheme. The Airborne Command Post was also based at Offutt AFB, Nebraska, and was always commanded by a general officer. This general could become the commander of all SAC forces if both the primary and alternate command posts were out of action. If this occurred, this general would receive orders from the president and Joints Chiefs of Staff. One of these aircraft would be on station, generally over the American Midwest, for around an eight-hour shift until relieved by the next aircraft and its crew. Should a replacement be delayed, the on-duty aircraft would remain, extending via aerial refueling if need be. This redundancy virtually assured availability of effective command and control of the U.S. nuclear deterrent force under any circumstances.[21]

The following aircraft provided the manned bomber and aerial refueling tanker capability during the period of the Cold War from the late 1950s. Versions of these aircraft are still in service with the Air Force.

Aircraft	Range	Speed	Weapons
B-52	7250 nm	456 knots	4 1 MT
KC-135	1300 nm	460 knots	

Table of mid and later Cold War bomber and tanker aircraft capabilities.

Boeing B-52

The design that would evolve into the B-52 started life configured with 4 large turbo-prop engines. The flying prototypes of the 8-engine jet-powered B-52 had the pilot and copilot seated in tandem seating. The first production models were the first to have the side-by-side cockpit arrangement that has persisted ever since. The B-52A was introduced into service in 1955. Approximately 80 B-52H models are still in service. A total of 744 aircraft were built between 1952 and 1962.

Powered by 8 Pratt and Whitney J57 turbojets, to cruise at 456 knots, at up to 47,000 feet, to carry up to 43,000 lbs. of bombs.

Range: 7250 nm

Boeing KC-135

Boeing developed the design of the Dash 80 with their own funds to offer the government a viable replacement for the KC-97, which created difficulties for the refueling of the B-47 and B-52 jet bombers. The bombers had to come down considerably in altitude and reduce speed significantly to refuel from the KC-97 propeller-driven airplane. Following initiation of the KC-135 production program, Boeing then created the 707 airliner prototype from the Dash 80.

Powered by 4 Pratt and Whitney J57 turbojets, to cruise at 480 knots, at up to 41,000 feet, to carry up to 33,800 gallons of jet fuel.

Range: 4350 nm

At the time of the Cuban Missile Crisis in late 1962, the U.S. strategic force of deterrence had become a mixed force. It was comprised of six hundred B-52s, seven hundred B-47s, nine hundred air refueling tankers, and one hundred seventy ICBMs. The location of bombers making up this force is depicted on the following map.[22] The map near the end of this chapter shows missile deployments at this same time.

Additionally, U.S. bombers could be deployed to "reflex" bases in Alaska, England, the Pacific, and the Mediterranean area to provide a shorter range to targets in the Soviet

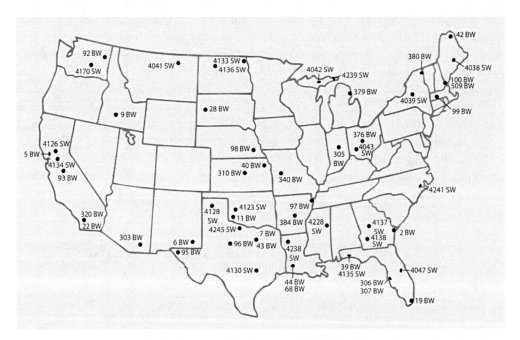

Map of Strategic Air Command bomber bases in late 1962.

Union. Primarily during the period of the B-47, aircraft were forward deployed complete with weapons and crew, to a base closer to the Soviet Union. These bases were Eielson in Alaska, Anderson on Guam, Sidi Slimane in Morocco, Brize Norton, Fairford, Lakenheath, and Upper Heyford in the United Kingdom, and Torrejon in Spain. The aircraft and crew would fly from their base in the U.S. to these forward locations, and normally remain at the forward base for ninety days, with periodic times at alert readiness.

EVOLUTION IN NAVIGATION

The need to fly long distances, virtually any time, to any location over wide areas of the earth, necessitated adopting new methods of navigation. The dual requirements that it must work the world over and not rely on facilities controlled by an adversary, or be subject to outage by the conflict, favored the use of the stars and planets or the earth itself as the reference, the two constants through a potentially global conflict. The former led to the refinement of celestial navigation and the latter led to the wider use of radar for navigation, both methods to determine relative position from the constantly available references. Each of these approaches measured the vehicle's position relative to the fixed reference, and by a succession of measurements, determined the rate and direction of the vehicle motion. Other military aircraft applications evolved Doppler navigation where a set of three or four radar beams were constantly monitored to determine the rate of vehicle motion over the ground, again measuring the motion relative to the constant earth.

With the advent of ballistic missiles, there was a need for improved targeting. With the manned bomber of World War II, the crew could be tasked to attack the railyards, port facilities, airfields, or factory complexes. Essentially, the bombs were delivered based on visual identification of the aiming point by the bombardier through an optical instrument for this purpose, the Norden bombsight. If specific location information was lacking, the crew could interpret the scene from the aircraft, known as pilotage, and drop the weapons based on this judgment. Such was not available to the unmanned ballistic missile; target location had to be established in a coordinate referenced scheme so that an automatic control system could properly steer the missile.

With the ballistic missiles as well as the submerged submarine, the earlier methods used by pilot and navigator crew members were not applicable or were impractical. The "stable platform" was invented to create an artificial reference from which to measure motion, or said another way, to measure the effects on the stable platform reference of the motion of the vehicle. The stable platform was a set of free-spinning gyroscopes mounted so that rotational motion about any of the three axes could be measured and accounted for. Adding to this platform three accelerometers to measure acceleration in the three axes accounts for all possible motion of the platform. Any rolling motion is sensed by the gyroscopes, and translation is sensed by the accelerometers. It is not incorrect to think of this as sensing the movement by the accelerometers, and knowing the direction to apply the corrections from the gyroscopes. The third element of this self-contained system is the careful keeping track of the history of these values and combining the sensed change in direction or position with the previously determined position to estimate a newer position. This mathematical process of determining a newer position based on the latest sensed changes is called dead reckoning. This dead reckoning process has been used for some time, first by ocean ships' navigators and later by the navigators in long-range aircraft.[23] The evolution that was needed for the long-range missiles and submarines, and later aircraft as well, was increased precision needed in order to accurately deliver weapons based on these vehicle positions.

STAND-OFF WEAPONS

With the increasing extent and capabilities of the Soviet air defense system, in 1956 the U.S. Air Force initiated development of a weapon that could be launched ahead of U.S. bombers to clear a path through Soviet defenses. The development was completed in only 30 months by North American Aviation, due to the similarity of the design with the previous Navajo missile by that company. The Air Force acquired 722 of these Hound Dog missiles before production ended in 1963. These were called stand-off weapons because they could be launched from outside a defended area, and thus significantly reduce exposure of the launching crew and aircraft to the defenses.

In 1960 these missiles, which have come to be called cruise missiles, began to be installed on B-52s. Most often five of the twelve B-52s on alert at a base were equipped with these missiles. The Hound Dog, or AGM-28, had a 400- to 785-mile range, depending on the flight profile used, and could cruise at 1,050 knots up to an altitude of 56,000 feet. Most often the flight profile was for the missile to initially descend after launch from the bomber, at perhaps 45,000 feet. Then after reaching a commanded pressure altitude, it would cruise seeking that pressure altitude until the on-board computer commanded it to end the cruise, from where it would enter its terminal maneuver of diving vertically onto the target.

The Hound Dog missile was powered by a J52 jet engine, also used by some jet aircraft. Following launch, the jet engine was run flat out at maximum power. Although called a missile, the vehicle had a delta wing and movable control surfaces. This combination really made the vehicle an unmanned aircraft.

The navigation system of the missile was a self-contained inertial system that included an astro-tracker, which remained with the launching airplane following launch of the missile. The astro-tracker would track a particular star, possibly the sun, to align the flight path. The angle where the optical sensor located the celestial object would allow determination of the aircraft's heading, which was transferred to the missile's reference as the basis for the inertial navigation of the missile.[24]

The Hound Dog missiles were part of the SAC alert force from 1961 until 1976, when they began to be replaced by the Short Range Attack Missile. These SRAMs largely filled the same role for the attacking force.

AIRBORNE ALERT

As the predictions of Soviet ICBM capabilities continued to grow, SAC sought to further assure a retaliatory response in the event of attack on the U.S. bomber bases.

In addition to the ground-based alert aircraft, trials were begun to develop the procedures to maintain some bomber aircraft on airborne alert. Airborne alert meant fully loaded bombers in the air with all materials onboard to conduct an attack, if ordered. The resulting flights were typically of twenty-plus hours duration, and required air-to-air refueling. These flights placed the B-52s on a path that took them to the northernmost extensions of North America, perhaps more than halfway from their base to their target. From these tracks, if the order came to proceed to their targets, they would discontinue the orbit track and proceed to the assigned target. There would be multiple B-52s on the same orbit track so that as one plane dropped out of the pattern on return to base, another plane would be joining the sequence, maintaining a constant percentage of the total force not at their base but ready to proceed to the target.[25]

East Coast bases generally provided aircraft into orbit tracks across Greenland or around the Mediterranean, while tracks of West Coast units took them north of Alaska. The circuits were arranged to pass near bases where the needed KC-135 refueling tankers were stationed to make it a bit easier to meet the bombers.

At each base involved, this required establishing an airline-type schedule of operations to prepare both aircraft and crews to depart at a specific time to maintain aircraft in the orbit scheme. Because of the long duration of the flights, special meals were prepared to sustain the crews in flight. Other changes included provisioning additional spare parts to allow the maintenance crews to keep the aircraft ready to go. The aircraft went through a more rigorous inspection after several flights to monitor any longer-term conditions. These activities constituted additional workload for the units involved; they still had routine training and ground alert schedules to maintain in order to assure their readiness on a continuing basis.

The trials for these operations began in 1959. Although there was more than one scheme used in the early operations, initiated during the Cuban Missile Crisis of 1962, by

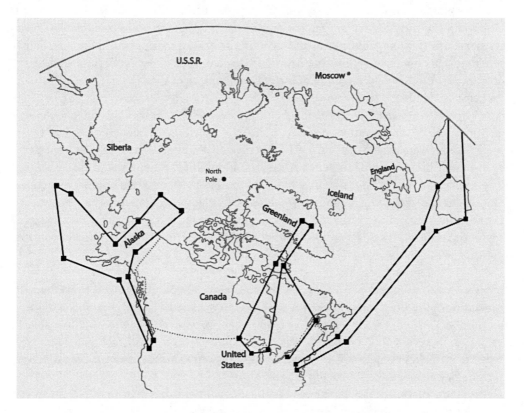

Map of Strategic Air Command airborne alert routes.

1966 the flights had settled into three routes: the northern route, where bases in the north and east U.S. provided four flights a day that looped across Greenland and back; the western route, where west coast bases generated two flights a day to cross Alaska and go further north before re-crossing Alaska and rounding the Aleutian Islands before returning to the lower forty-eight; and the southern route where east coast bases would equip six flights a day to cross the Atlantic and loop through the Mediterranean by crossing Spain to return across the ocean to their east coast origin. These twelve flights a day, sometimes called the daily dozen, became normal routine during this time.

These airborne alert flights were constant until 1968. Although quite reliable and safe by almost any standard, two accidents that left nuclear cleanup tasks in Greenland and near Spain helped convince leadership that the flights were no longer required. The growth of the submarine-based deterrent force doubtless assisted in this decision to discontinue.

Evolution of the B-52 Bombing Navigation System

The development of the bombing navigation equipment of the Cold War began even as the Norden bombsight, described previously, was being used to attack Germany and Japan. Using radar to locate the target and to determine distance from the target was the logical next step.

The bombing problem, as described for the Norden case, was unchanged for a manned bomber in the early Cold War era; an unguided free-fall bomb was slowed down by air resistance after leaving the airplane, causing it to land behind the position of the continuing airplane. This distance behind the dropping airplane was called Trail, and had to be accounted for in the determination of the release point, as did the Actual Time of Fall for the selected bomb.

What was different was the planned altitude from which the bomb would be released and the speed of the aircraft used. Early in this relatively long development period, these specifications were 60,000 feet and 650 knots respectively. By 1950 these values were essentially doubled as next generation bombers, beyond the B-52, were in the planning stages. The B-52 was to bomb from as high as 30,000 feet at a speed of over 600 knots. This greater altitude increased the time the bomb would be subjected to winds, increasing the uncertainty of the hit, and increased the need for the most precise release point possible as a response. Additionally, this increased altitude also seriously limited the effectiveness of the optical sight as a sensing element of the target location.

Coupling a radar capable of searching out the target and tracking it when within 15 miles with a computer to account for the bomb trail distance and time of fall was the first generation of these electronic systems. Such a system was used on the B-50 and B-36 strategic bombers. This was another early example of an integrated system of multiple "black boxes," each performing a subset of the total task to be done, and each provided by a separate manufacturer, specializing in that part of the problem. For instance, the radar was provided by a company specializing in radars.

As the evolution of this equipment continued, in order to equip the B-47 jet, the range for radar tracking of the target had to be increased to fifty miles to account for the higher-speed aircraft. Additionally, the concept of the offset aiming point was introduced. An offset aim point is a point in the vicinity of the actual target which could be more easily identified by the bombardier with the system, in this case by the radar return. The system then translated the weapons release point by the selected offset relationship from the radar-tracked aim point.[26]

The intercontinental range of the SAC bomber force, and the principal areas of targets requiring navigation near the North Pole, where magnetic instruments had limitations, established the requirement for a method to determine earth orientation from the stars. This requirement led to the astro-compass. This instrument implemented the methods of celestial navigation by tracking the location of stars, or the sun, via optical telescope, and calculating a reference to true north, which was provided to the larger system. The operator would set the optical telescope on the star to be observed, then initiate the tracking. The star and its celestial parameters, extracted from the Air Almanac, were provided through an input panel enabling the instrument to perform the calculations resulting in the determination of true north.

Similarly, the need to have accurate sensing of the aircraft's motion over the ground led to the addition of the Doppler radar sensor to the larger system. By generating multiple radar beams at specific directions from the aircraft to the ground, and measuring the individual motions of the ground relative to each radar beam through the Doppler principal, the complete motion of the aircraft over the ground could be computed. This allowed extraction from the complete motion of the wind components, to be provided to the navigation computer.

The addition of these sensors to the system occurred in the early models of the B-52, the A, B, and C versions of the aircraft.

The bombing navigation system of the B-52D, E, F, G, and H, i.e., all active B-52s between the years 1964 and 1980, was the ASQ-38 Offensive Weapon Control System. This system was made up of the individual instruments, or black boxes, of search radar, automatic astro-compass, true heading computer system, Doppler radar, and bombing/navigation computer.

The search radar allowed the operator to search for a particular feature in the terrain, such as a coastline, river inlet, mountain, or structure such as a bridge or prominent building. Such a feature could be tracked automatically by the radar once selected, allowing continuous values of distance, and azimuth and vertical angles to be provided to the system. Additionally, this tracking capability of the radar freed the bombardier to monitor and control the other system elements as well.

The astro-compass tracked up to three stars in order to determine the direction of true north.

The Doppler radar measured the aircraft's motion over the Earth's surface and provided values of ground speed, ground track angle, wind direction and speed, and values of the aircraft's position based on the other values being applied to a prior position.

The bombing or ballistics computer incorporated all the values from these system sensors, applied the aerodynamic effects of a particular bomb's path, and determined the appropriate time for weapon release.

By the 1960s, the formation tactics previously presented for the World War II B-17 aircraft had been updated for application of the B-47 and then the B-52 aircraft. In addition to the trend of higher and faster flights to remain free of defenses, the capacity to defeat the electronic aids that were integral to those defenses was a substantial difference in military aircraft after World War II.

The basic formation of the B-52s is three airplanes with a spacing of one mile between planes along the flight path. Vertically they are each separated by 500 feet of altitude with the lead aircraft at the low position and each succeeding aircraft stepped up 500 feet.

Laterally, the following planes align so that their wingtips would essentially touch the lead plane if not for the trail spacing. An additional cell or cells could be placed in trail to bring more bombs onto a particular target.

Just as the previously described bomber aircraft formation of World War II was optimized to maximize the effect of the defensive guns, B-52 formation was optimized to obtain maximum benefit from the onboard electronic jamming equipment. The jamming pattern achieved by the B-52 antennas was an area forward of the aircraft.

The onboard ECM system was operated by the electronic warfare officer of the B-52 crew. Through his sensing equipment he would identify the signals from the type of enemy equipment that threatened his aircraft, and others, and "set on" the

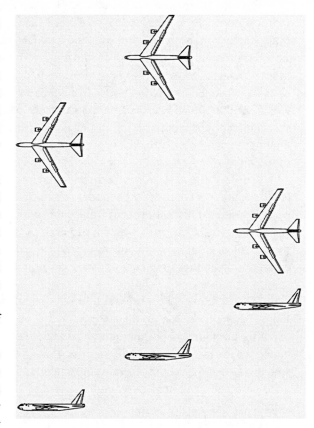

B-52 formation as seen from above and to the side.

jamming transmitters of the aircraft to oppose the threatening signals as best he could.

1970s Example

Although the B-52s have never been called upon to perform their nuclear delivery missions, they were used during the Vietnam War, between 1965 and 1972. The following illustrates the use of the B-52 against strategic military targets in North Vietnam. These operations were conducted in late 1972 as part of Operation Linebacker II. The integrated air defense system, described in a previous chapter, in many respects illustrates the same elements present in the air defenses of North Vietnam in 1972. These defenses were made up of Soviet-made equipment and the crews trained by Soviet advisors. As a result, it is not a stretch to project that an important target in the Soviet Union would have been protected by a similar scheme.

The lack of military equipment factories in North Vietnam made the railroads, power facilities, communications, and related sites the principal strategic targets in that country. The rail lines, bridges, marshaling yards, transshipment yards, and warehouse districts were the locations of the weapons and other equipment received on the two main rail conduits from China. The same types of infrastructure supporting the port at Hai Phong would have had the same priority if not for the mining of the harbor that the U.S. had carried out at the start of this period of escalated conflict. Additionally, the power generation capacity of the country was the second priority for the bombing attacks.

The defense system of North Vietnam consisted of long-range search radars, shorter-range ground control intercept (GCI) radars, surface-to-air missile (SAM) interceptors, antiaircraft artillery (AAA) guns, and jet aircraft interceptors utilizing both guns and air-to-air missiles. The SA-2 surface-to-air missile incorporated both a target tracking radar and a missile guidance radar. These sensing and response elements were tied together by a central command and control arrangement. All of these defensive radars constituted a formidable cast of electronic signals to be neutralized in order for U.S. aircraft to successfully operate in the area. These air defense installations were targeted by attacks to reduce losses among the American air elements.

By the 1970s, there were several approaches to neutralizing the electronics systems of a defense installation. Much of this comes down to use of radio frequencies, or bandwidth. The broader topic is electronic warfare, defined as "the use of electromagnetic energy to determine, exploit, reduce, or prevent hostile use of the electromagnetic spectrum and action that retains friendly use of electromagnetic spectrum." In short, the U.S. wants to be able to use its radar and communications while preventing an enemy from doing so with their equipment.

The U.S. had evolved people, equipment, and tactics to oppose enemy defensive electronic systems; this capability was termed defense suppression. It could take two forms: short-term interruption of the system, or destruction of the system through attack. Short-term interruption was most often called jamming, of which there were both active and passive types. All three types of suppression were used in the American attacks against North Vietnam in 1972. All bombing aircraft participating in these missions carried active electronic jamming equipment. Additionally, there were supporting aircraft in the vicinity to provide specifically targeted jamming, and other aircraft to distribute chaff—a passive material designed to reflect enemy radar signals and thereby cause confusion. Also, there were aircraft to seek out and destroy enemy SAM sites. These activities are all termed Electronic Countermeasures (ECM).

Active jamming is the transmitting of an electrical signal on the same frequency as used by another, for the purpose of interfering with the activities of that other user. American bombing aircraft carried jamming equipment to make the enemy radar ineffective in the immediate vicinity of the American aircraft. Other dedicated jamming aircraft would take up an orbit position some miles from a target area and project jamming signals across the area containing the enemy defenses needing to be neutralized. Similarly, the chaff-laying aircraft would fly in a formation to create a "chaff cloud" perhaps five to ten miles wide and thirty miles long, which the bomber aircraft would hide behind or in, attempting to keep at least part of the chaff between themselves and the enemy radar sites.

The supporting aircraft with equipment to seek out and destroy the enemy SAM sites would patrol the area monitoring for the specific radar transmissions that their equipment was designed for. These Wild Weasels, on the mission of Suppression of Enemy Air Defense (SEAD), could launch a missile that would home in on the transmitter of the specific signal. Similarly, aircraft equipped to attack and destroy enemy fighter aircraft would perform a combat air patrol (CAP) along the corridors the bombing aircraft used on entry and exit from the target area. These CAP aircraft would seek out and destroy, normally with air-to-air missiles, any enemy aircraft that might threaten any of the friendly aircraft mentioned. Airborne Early Warning (AEW) aircraft, essentially an air mobile radar station, would orbit just beyond the range of enemy missiles to provide warnings and direction to all American aircraft in the area.

These aircraft supporting the flights of strategic bombers attacking a group of targets could easily number over one hundred. The individual mission illustrated was comprised of 120 B-52s. These numbers do not include the air refueling tanker aircraft required for these operations. There were two refueling orbits over the Gulf of Tonkin and nine more over Laos. Each of these orbit tracks would nominally contain six KC-135 tankers. Lastly, search and rescue aircraft, in the form of rescue helicopters and propeller-driven airplanes capable of remaining in an area for an extended time, would swing into action should any aircrew of any of the aircraft types mentioned find themselves down in hostile territory for any reason.

The following graphic illustration shows Hanoi, Haiphong, and the surrounding areas of North Vietnam. These population centers are located in the fertile valley of the Red River delta, where the Red and Black Rivers empty into the South China Sea.[27] The hatched circles show the location and coverage of the SAM sites; the solid lines show the routes of the B-52 bombers inbound to their targets; the lines with small dashes show the exit routes; and the wider dashed lines show the orbit stations of supporting aircraft. Targets in the immediate vicinity of Hanoi were protected by the more than 15 SAM batteries depicted. Although this illustration attempts to show the SAM positions at a given time, it also could be wrong with respect to any individual site. A site could and would be relocated from the previous day's location to confuse the attackers or to fine tune the coverage for a particular target or route.

The bombing target would be identified by the radar navigator of the B-52 crew. Most often this navigator would have selected a prominent point that would provide a distinctive return on the airplane's radar, and established the actual target for the bombs as a range and bearing from this point of radar return. This "offset point" was set up in the bomb-nav system, which would release the weapons when this relationship was achieved. In Hanoi, this ground object to provide a distinctive radar return was often the main high-way/rail bridge crossing the river near the center of the city.

The foregoing described the methods that the U.S. Air Force, with Navy support, used to conduct an air campaign against strategic targets in North Vietnam in 1972. What it was not was the massive nuclear retaliatory attack that would be launched by the U.S. against the Soviet Union and China following detection of attack on North America. It should be considered a valid representation of the tactics that would have been employed if such an attack had been undertaken at the time. The B-52 aircrews participating were transferred from their stateside bases to either Guam or Thailand with their airplanes. At the time of their first mission from, say, Guam, it may have been only a few days from their most recent period of standing nuclear alert at a stateside base. These missions were the closest any

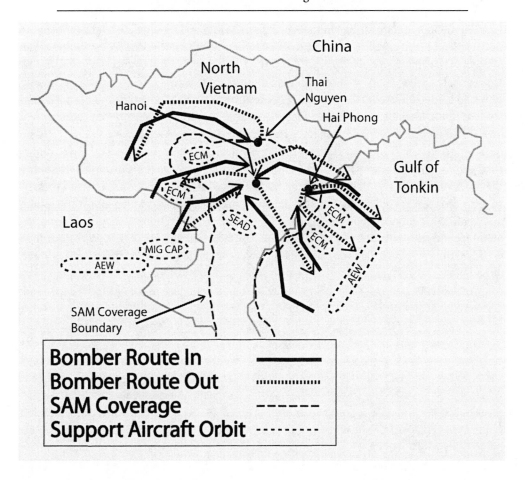

Map of portion of North Vietnam showing B-52 flight routes, SAM defenses and orbit points for aircraft supporting an attack.

SAC crews ever came to executing their long-prepared-for nuclear attack missions against targets in the Soviet Union.

This series of attacks was successful for the Americans in that they caused the North Vietnamese to seek the restart of peace negotiations, which ultimately ended the hostilities. However, the operations were not without losses. There were over 1000 SAMs launched by the defensive forces against the attackers in the twelve-day campaign. Ten B-52 bombers, three F-4 of various specialties, two F-111, three A-7, three A-6, one HH-53, and one RA-5C aircraft were lost to the defenses. Forty-one American air crewmen were killed and another 45 became prisoners of war.

The First Long-Range Missiles: Thor, Jupiter and Atlas

Although military missiles trace their evolution to China in the twelfth century A.D., there was very little European activity on this subject until the 1800s. Then rockets were used in the War of 1812 by the British, by the Americans in the Mexican War, and by both

Union and Confederate armies in the American Civil War. These uses of rocketry were not decisive, and this usage was discontinued by the end of the 1800s due to improvements in artillery. Research and development continued in a number of countries, including the United States, where Robert Goddard developed several experimental rockets in the early 1900s. In 1918, Goddard, after proposing rockets for both the U.S. Army and Navy, demonstrated tube-launched, solid-fueled rockets suitable for either ground or air launch. With World War I ending just five days after this demonstration, the military's interest did not develop. Goddard, however, did continue his research, but without substantial funding from the military. In the 1930s, Theodore von Karman encouraged a group of students at the California Institute of Technology to pursue their interest in rockets. This led to Army support of research to develop a rocket capable of assisting a heavy aircraft off the ground at takeoff. This Caltech group developed a number of useful fuels and motors from this work, but this did not lead to any widespread application. During World War II both Dr. Goddard and the Caltech group worked on rocket-assisted takeoff of conventional aircraft, although the subject had been renamed Jet Assisted Take Off, or JATO.

In the 1930s it was a German group that seized upon the ideas of Dr. Goddard, and combining them with ideas from Herman Oberth, proceeded to develop military rockets for Germany. Under the Versailles Treaty ending World War I, Germany was prevented from developing conventional weapons, so used the unrestricted rocket field to develop longer-range artillery. In 1937–38, after developing the experimental test area at Peenemunde on the Baltic Sea, the Germans flew their A-4 rocket. The A-4 was 46 feet high, 5 feet in diameter, weighed 14 tons, and could transport the 1,650-lb. warhead approximately 200 miles. Because the test program did not produce constant positive results, and Germany had other priorities at the time, Hitler did not approve the rocket for production until 1943, when it became known as the V-2 rocket. The "V" name, which stood for vengeance, had been applied by Joseph Goebbels, the propaganda chief of the Nazi regime.[28] Even after the V-2 had been approved for production, the Germans did not use these weapons until they had acquired a supply of them. Between September 1944 and March 1945, Germany launched 3,745 of these V-2 rockets. Of these, something over 1,000 were launched at England.

American interest in the V-2 rocket grew slowly as a result of wartime intelligence gathering, which uncovered the German activity to construct the unmanned missile. During the attacks on England, remnants of the missiles were recovered and examined. At the end of the war, examples of the V-1 and V-2, along with scientists and technicians from the German developments, were transferred to the U.S. where the Army continued testing and evaluation of the German designs.[29]

A number of activities in the U.S. were slowly advancing the state of missile development and capability. More than one U.S. aircraft company saw the signs that termination of the large World War II aircraft production contracts, with the close of hostilities, meant they would have to find work in other areas in order to survive. Consolidated Vultee, commonly known as Convair, Northrop, and North American Aviation all established missile divisions and took on the work of the early development of unmanned bomb-carrying vehicles. At this time there were strong supporters for both winged air-breathing vehicles and non-winged rocket-powered momentum vehicles. The air-breathing winged vehicle was essentially a pilotless airplane; this vehicle configuration eventually came to be termed the cruise missile. The rocket-powered concepts differed in that drawing oxygen from the air was not required as the fuel contained oxygen to support combustion, thus allowing that vehicle to exit the atmosphere. These became known as momentum or ballistic vehicles because once the fuel ran out, the vehicle would continue in the direction established during the powered portion of the flight and return to the ground following a ballistic arc. When numerous projects were canceled by the postwar budget reductions of the Truman administration, it was mainly the air-breathing pilotless aircraft that continued in funded development. It was thought these projects would produce deployable equipment sooner. Both Convair and North American continued some ballistic missile research work on their own funds, believing that the U.S. would eventually have to have missile capability.

With the Soviet testing of an atomic weapon in 1949 and the start of the Korean War in 1950, funds did become available for missile developments. However, between 1946 and the mid–1950s, there was intense rivalry among the branches of the U.S. military, after a 1947 organization that provided for the Navy Department, and Departments of the Army and the new Air Force. Through the mid–1950s all three services developed missiles with overlapping capabilities. Despite several high-level committees, study groups, and recommendations of special advisors, the new Department of Defense and the Joints Chiefs of Staff failed to clearly delineate areas of responsibility with regard to missile development and operation. This internal battle, over what is often called the "roles and missions" definitions, allowed duplication of work by the service branches in the areas of surface-to-surface missiles, surface-to-air missiles, and air-to-air missiles.

It was the intelligence assessments of the capabilities of rockets being developed by the Soviet Union that jarred the U.S. into establishing a program to develop and field an Intercontinental Ballistic Missile, or ICBM. This area of long-range strategic missiles was the least disputed by the other service branches, and the Air Force had only internal debate to establish specific requirements for an operational system. The principal subject needing agreement was the weight of the atomic warhead payload. The requirements for range and accuracy were much less disputed, but the weight of the payload to be carried the 5500 or so miles had a tremendous impact on the proposed configuration of the rocket vehicle. The earliest design concepts assumed a 7000-lb. payload, whereas following the Castle nuclear weapons tests in the Pacific in 1954, a payload of 3000 lbs. was understood to be feasible. This greatly shrunk the size of the required rocket system.

These internal debates established the requirements of the Atlas program, which called for a missile to deliver a 3000-lb. warhead 5500 miles and hit the target within 1 mile. Convair was contracted initially for design studies of missile topics while these specifications and other matters were finalized.[30] The development time of the required rocket engines generally set the pace for any test flights. The frequent difficulties with budget in this post–World War II period caused much cycling among different concepts and quantity of missiles to be deployed by the highest levels of leadership. The crucial component research and development of the key subsystems continued in the background. Additionally, the risk of a failure in the then single ICBM program brought about a parallel development path for a second missile type, largely with contractors independent of the Atlas development. It was thought that these alternate subsystem components (propulsion, two types of guidance, nose cone, and computer) might be backup for failures in the primary component developments and that the alternate missile was backup for the entire Atlas system. This second ICBM came to be the Titan missile, as proposed by the Glenn L. Martin Company.

The perception of the need to field missiles with atomic warheads prior to a similar achievement by the Soviets not only fueled this parallel path for an ICBM, but initiated new requirements for an Intermediate Range Ballistic Missile (IRBM), which could be operational at an earlier date. The IRBM is considered to have a range of 1000 to 2000 miles, whereas the ICBM is considered to have a range of 5000 miles or more. A partial instigation for another new weapon system was the request by longtime ally Britain for a missile that could be based in that country to provide deterrence against the Soviet Union. The commander of U.S. forces in Europe also stated the requirement for missiles to be based in Europe. This caused the Air Force to respond by requesting other aerospace companies, so as to not draw resources from the ICBM work, to propose solutions. This selection process led to the Thor missile, as proposed by the Douglas Aircraft Company.

This new requirement for a missile of intermediate range with a sooner deployment date than then envisioned for the ICBM resurfaced the "roles and missions" debate within the U.S. service branches. The Army, supported by the Navy, was developing its own missile of generally these same capabilities known as the Jupiter. This missile development was being overseen by the former German rocket team headed by Wernher Von Braun, now established as an Army agency at the Redstone Arsenal located in Alabama. The three Air Force missile developments were overseen by the Air Force's Western Development Division headed by Bernard Schriever, at the time a major general. This Western Development Division had evolved from a special project organization for the initial ICBM (Atlas) development to become the Ballistic Missile Division by 1957. Early in 1956, the four missiles, two ICBMs (Atlas and Titan), and two IRBMs (Thor and Jupiter), were under contract for development by the United States.[31]

Although the ICBMs, and then the Air Force and the Army IRBMs, were at the highest levels of priority for development by the U.S. defense establishment, they were still subject to President Eisenhower's desire to reduce defense spending. Although this did not affect the technical research and development work on the subsystems, it led to much cycling in the plans for how many, what type, and at what rate to deploy the missiles. The required reaction time to launch these weapons also affected the number of people and the amount of equipment that would have to be available in the new missile force. The requirement established during this period was that each unit would have to launch 25 percent of its force within 15 minutes of an alert, an additional 25 percent within two hours, and the remainder of the force within four hours.[32]

Adding to the cost and complexity of how to meet these launch requirements were the deployment configurations, which included the number of launchers, launch control centers, and related equipment. This was an important consideration for the liquid-fueled missiles, whose fuel tanks were loaded with the volatile fuel components immediately prior to launch, and for the earlier missiles the rocket was then rotated or lifted into the firing position. The number of launch positions determined the number of sets of this critical support equipment that had to be available, hence heavily influencing the rate at which individual missiles could be made ready for launch. In turn, the number of launch positions was certainly affected by the basing scheme. Of these early missiles, only the Titan was initially intended to be stored in a "hardened" underground silo location that included an elevator to lift the missile to firing position from its underground storage location. The Atlas, Thor, and Jupiter missiles initially were planned for being stored in the vertical position, in the open without physical protection. For these missiles particularly, protection of these sites against enemy attack was largely the distance between sites, or the dispersion distance. This distance was set so that an attack on one site would not affect the adjacent site. This forced an attacker to apply many more weapons to destroy the entire missile force. But all of these considerations affected the construction costs and the purchasing costs of the required equipment and were the subject of much scrutiny in the budget-conscious administration of the late 1950s. As a result, particularly in 1957, the Air Force generated multiple proposals for the initial deployment, primarily to show the near-term cost effects of different rates of production and the rate at which missile sites would become active. Based on National Security Council recommendation, the Air Force had established the need date of January 1959 for initial operational capability (IOC) based on the perception of when the Soviets might be capable of a first strike. During this time, the required total missile force to be built was reduced to lower the rate of expenditure as well. The various proposals by the Air Force detailed different paths to achieve the initial operational date, and varied in the rate that the remainder of the total planned force would become available.

The launching by the Soviets of the Sputnik satellites in late 1957 led to an increase in the priority and spending for efforts by the U.S. to establish a ballistic missile force despite the desire to decrease spending on defense during these postwar years. The Sputnik satellites made it clear to all that the Soviet Union had an advanced rocket program that could deliver nuclear warheads to almost any place on earth. This apparent lead in the launch capability by the Soviets was labeled the "missile gap" by political opponents of the Eisenhower administration. The missile gap argument had it that the U.S. would lag the Soviets in ICBM ability until perhaps 1963, and this weakness by the U.S. created an opportunity for the Soviets to launch an attack to which the nation could not respond in kind.

What resulted from all the planning for deployment of America's first ICBM force was a varied set of configurations. The deployment decisions were influenced by the desire to establish some deterrence against a Soviet threat as early as possible and the desire for practicality, reliability, and economy. The first two Atlas D squadrons were of six missiles each, configured with three missiles per launch complex. The next two Atlas D squadrons were of nine missiles each, configured with three missiles per launch complex. The next three Atlas E squadrons were of nine missiles each, configured with one missile per launch complex. The final six Atlas F squadrons were of twelve missiles each, configured with one missile per launch complex. For protection, the Atlas launch complexes were dispersed eighteen miles from each other. The Atlas D sites were not hardened, the Atlas E sites were semi-hardened to withstand a blast of 25 pounds per square inch (psi) pressure, while the Atlas F sites consisted of underground silos hardened to 100 psi. The first Atlas D site became operational in January 1959, and the final Atlas F squadron became operational in December 1962.

All six squadrons of the Titan (I) missiles were of nine missiles each, configured with three missiles per launch complex. The Titans were deployed with each missile in an underground silo hardened to 150 psi. The initial Titan design, with the volatile fuel that could only be loaded aboard the missile shortly before launch, became known as the Titan I. Design improvements, which would allow for long-term storage of the fuel in the missile and launch from the underground position, became known as the Titan II. The first Titan

I squadron became operational in April 1962 and the final Titan I unit went on alert in September 1963. For all the above configurations, the launch complex contained the launch control center where the launch crew initiated a launch.[33]

In addition to cost and political realities that the Atlas and Titan ICBM deployments faced, the Thor and Jupiter IRBMs also faced some difficulties owing to the concept of deploying these missiles in other countries. It was known for some time that Thors would be based in England at the request of the British government. The plans to base three Jupiter squadrons in France were rejected by newly elected French President Charles DeGaulle in 1958. Additional plans to base Thors in Italy, Turkey, Okinawa, and Alaska never got out of the proposal phase, but following the French rejection, agreements were reached to base the Jupiter in Italy and Turkey. As a result, four squadrons of Thors were based in England, two squadrons of Jupiters were based in Italy, and one squadron of Jupiters was based in Turkey. All seven IRBM squadrons were of fifteen missiles each, configured with five missiles per launch complex, which contained the launch control center.[34]

Air Vehicle	Range	Weapons
Thor	1300 nm	1.44 MT
Jupiter	1300 nm	1.45 MT
Atlas	4800+ nm	1.44 MT, 3.75 MT
Titan I	5500 nm	3.75 MT
Titan II	7800 nm	9 MT
Polaris	1000, 1500, 2500 nm	200 KT, 200 KT, three 200KT
Minuteman	7000+ nm	1.2 MT

Table of U.S. ballistic missile capabilities.

THOR

The Thor missile was an inertially guided intermediate-range ballistic missile. The missile was protected by a long tubular building that moved back on rails for erection and launch. The missile was stored horizontally until the covering building was moved, then rotated 90° to an upright position. When the missile was upright, the fuel was pumped into missile tanks. As a result the preparation for launch took about 15 minutes from the time an alert order was received. The Thor was deployed three missiles per launch complex, and five complexes made up a squadron of fifteen missiles. The Thor had a range of 1500 miles and carried a single nuclear warhead of 1.44 megatons. Although the Thor did not last long as a nuclear-armed strategic missile, it became the first stage of the well-used rocket known as the Delta/Thor, the first of the Delta family of space launch vehicles. Newer versions of these rockets are used to place satellite payloads in Earth orbit and are still in use today.

JUPITER

The Jupiter missile was an inertially guided intermediate-range ballistic missile. The missile was stored vertically in the launch position. Making the missile ready for launch took about 15 minutes from the time an alert order was received. During this time, the fuel was pumped into the missile tanks and the inertial system was aligned. The Jupiter was deployed three missiles per launch complex, and five complexes made up a squadron of fifteen missiles. The Jupiter had a range of 1500 miles and carried a single nuclear warhead

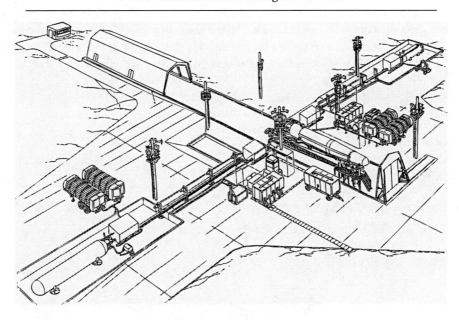

Depiction of a Thor missile launch site.

of 1.45 megatons. The launch crew for the Jupiter consisted of three members: a launch control officer and two crewmen.

ATLAS

The Atlas D used a radio-inertial guidance system. The missile transmitted its inertial measurements to the ground station. The ground station computed and transmitted back

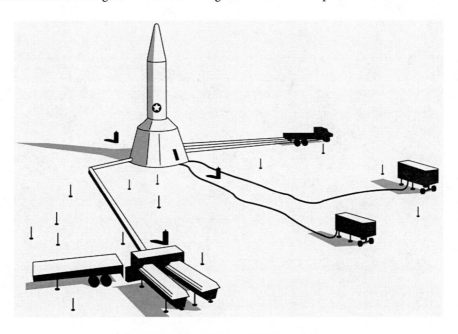

Depiction of a Jupiter missile launch site.

to the missile the course corrections. The Atlas D had a range of 5500+ miles and carried a single 1.44 megaton nuclear warhead. The Atlas Ds were stored in the open in the launch position. As with the other liquid-fueled rockets, the Atlas had its fuel pumped into the missile's tanks just before launch.

Launch crew of 12 for the D model:

Launch control officer
Missile system analyst
Power distribution system technician
Missile electrician
Missile maintenance technician (3)
Missile engine mechanic
Ground support equipment specialist
Propulsion system technician
Guidance system analyst
Hydraulics technician

The Atlas E used an all-inertial guidance system. The missile was stored horizontally under a covering building known as the coffin. The missile was rotated 90° to an upright position for loading of the fuel before launch. The Atlas E had a range of 6300+ miles and delivered a single 3.75 megaton nuclear warhead.

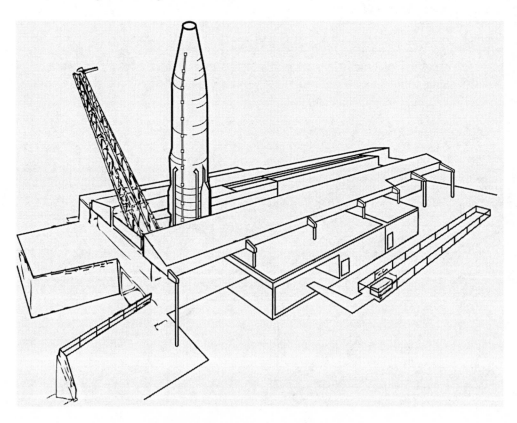

Depiction of an Atlas D missile launch site.

The Atlas F continued with the all-inertial guidance system, but the missile was stored vertically in an underground hardened silo. The missile's fuel could be loaded into the tanks while underground; one of the fuel components could be stored in the missile long-term. After final fueling, the missile was lifted to the surface by an elevator for firing. This allowed one Atlas missile to be launched every ten minutes, faster than for the D and E versions. The Atlas F had a range of 6300+ miles and delivered a single 3.75 megaton nuclear warhead.

Following retirement of the Atlas missiles as strategic missiles, all the Atlas D, E, and F rockets were used as space launch vehicles to place countless payloads in orbit. Atlas rockets were the launch vehicles for the last four launches of Project Mercury, which was the first U.S. program to place astronauts in Earth orbit. The Atlas rockets were used for over twenty years as generic space payload launch vehicles.

Launch crew of 5 for the E and F models:

Missile combat crew commander
Deputy missile combat crew commander
Ballistic missile analyst technician
Missile facilities technician
Electrical power production technician

Support—non-launch crew:

Guidance control technician
Engine technician

Eleven bases in eleven states were selected for the Atlas missiles. The first Atlas D launch site became operational at Vandenberg AFB in California in January 1959, and the final Atlas F squadron became active at Plattsburgh AFB, New York, in December of 1962. Between these dates, squadrons became active at bases in Wyoming, two bases in Nebraska, Washington, two bases in Kansas, Oklahoma, Texas, and New Mexico. These missile sites were inactivated between mid–1964 and early 1965. These Atlas sites were inactivated fairly early due to the improved reliability, simplicity, and lower overall operating costs of the newer Minuteman.

Titan I

The Titan I was a radio-inertial guided intercontinental ballistic missile. A central radar facility provides the guidance for three launch complex sites. Each missile was stored beneath the surface and then raised to the surface prior to fueling and launch. The liquid fuel had to be loaded into the chamber of the rocket motors just prior to launch. The Titan I had a range of 6300 miles and delivered a single 3.75 megaton nuclear warhead. The guidance radar was centrally located to three launch complexes as depicted below. One such operation had the radar located at Larson AFB near Moses Lake, Washington, and three missile sites located nearby near Schrag, Warden, and Frenchman Hills. The preparation for launch could be completed so that the second and third missiles could be launched at seven-and-a-half minute intervals following the launch of the first missile.

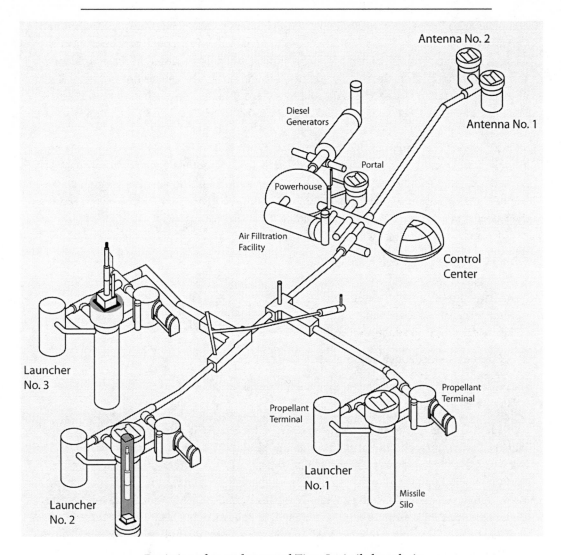

Depiction of an underground Titan I missile launch site.

Launch crew of 4:

Launch control officer
Guidance control officer
Ballistic missile analyst technician
Missile maintenance technician

Five bases in five western states were chosen to deploy the Titan I missiles. The missiles became operational in Colorado, Idaho, South Dakota, California, and Washington between April and September of 1962. Compared to the later missiles, these deployments had a relatively short lifetime, ending in January and February of 1965. The trio of shortcomings of this missile were that disabling the single guidance radar prevented launch of all nine missiles of a squadron; each missile had to be fueled prior to launch, resulting in a higher exposure to attack; and only one missile could be under the guidance of the centrally located radar facility, reducing the rate that all the missiles of a squadron could be launched. Additionally, after these missile sites were established, the Soviets had the capability to destroy U.S. targets at these ranges, creating a preference for U.S. sites further from Soviet territory.

TITAN II

The Titan II was an inertially guided intercontinental ballistic missile. A total of 54 missiles were spread among 3 wings and located at Davis-Monthan AFB near Tucson, Arizona; McConnell AFB, Kansas; and Little Rock AFB at Little Rock, Arkansas. A wing was comprised of 18 missiles. The underground silo and launch complex housed the missile, the launch control room, and spaces for the crew. Of the crew of four on a 24-hour alert shift, two were required to be in the control room. The other two could eat, sleep, etc., and could be in the crew quarters. Crew quarters had two bunks, a kitchen, and a table. Crews would cook for themselves. The crew of four commonly switched off two at a time. There was one control room and one crew for each missile. The crew completed a detailed checklist of the entire facility every 12 hours. Any discrepancies were called in to be scheduled for service. There were normally no security personnel on site. A radar alarm system protected the silo cover area and the air vent opening, which could alert the crew to activity at the ground level. The Titan II missile could carry its single 9-megaton warhead 9,000 miles. One innovation of this missile was that fuel could be stored in rocket tanks. Oxidizer had to be kept 60° F to not boil off. The missile's gyros were kept partially spun up to shorten the time to launch. The Titan II rocket was also used as a launch vehicle for other than military payloads. Notably the Titan II was adapted as the launch vehicle for the Gemini space program of the mid–1960s. All manned Gemini missions were launched into earth orbit by a Titan II booster.

Crew of four. Two were launch officers, one NCO maintenance, and one NCO analyst:

Missile combat crew commander
Deputy missile combat crew commander
Ballistic missile analyst technician
Missile facilities technician

Two squadrons each of the Titan II missiles were located at bases in Arizona, Kansas, and Arkansas between January and September 1962. These sites remained operational nearly to the end of the Cold War, being inactivated between 1984 and 1987. With its 9,000-mile range, these missiles could be sited in states further from the Soviet Union and still provide target coverage for all potential targets.

The Thor and Jupiter installations were operational until 1963 and were then discontinued as obsolete or just no longer necessary. The removal of the Jupiter missiles was an unpublicized (at the time) element to the agreement to remove Soviet IRBM missiles from Cuba as the conclusion to that crisis in 1962. The Atlas missiles remained operational until either 1964 or 1965, when their mission was taken over by the expanding deployment of the Minuteman.

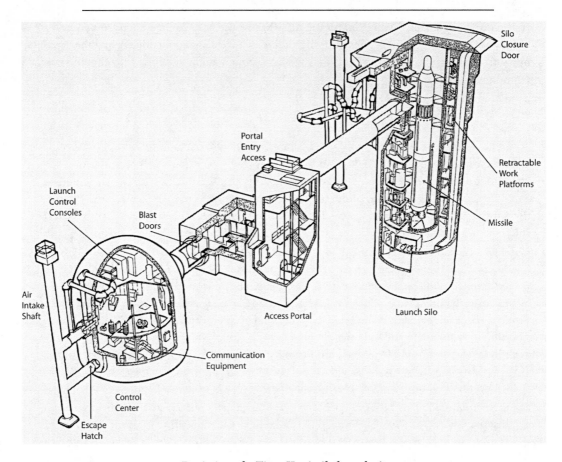

Depiction of a Titan II missile launch site.

Establishing Submarine Bases: Polaris, Poseidon and Trident

In the summer of 1955 the Killian Report had suggested that the U.S. place IRBMs on sea-based platforms. The National Security Council made this a recommendation, and the Navy was assigned to develop the necessary support structure to host the Jupiter missile on ships, then being developed by the Army with Navy support. The directive was to develop a 1500-mile surface-to-surface weapon system that would provide the Navy with an offensive capability against land targets at this range. Initially, it was not specified that the launching platform be a submarine, and in fact the resulting missile was first fired from a surface ship. Initial studies showed that a missile capable of boosting a 3000-lb. warhead the desired 1500-nm range was considerably larger than practical for shipboard operation. The Navy continued to study shipboard configurations with the ultimate objective of arriving at a compatible solution for that service.

As with the early stages of the Atlas ICBM program, the three technology areas establishing the size of the required missile were the focus of the R&D effort. First, in early 1956 the Navy had initiated a study of solid-propellant rocket propulsion for the Fleet Ballistic Missile, the generic name for what would become the submarine-launched ballistic missile. The resulting development of solid fuels reached the point where an IRBM could be configured without the liquid fuels of all the prior missile concepts. Second, the MIT Instrumentation Laboratory was investigating much smaller components for the missile's guidance system. Demonstration of these new components followed. And third, the Atomic Energy Commission, responsible for nuclear weapons development for all of these delivery vehicles, predicted even smaller nuclear warheads were possible. These three breakthroughs made a solid-fueled nuclear-armed missile launched from a submarine fully feasible.[35] When the combination of much more capable solid fuel, a much smaller and lighter guidance system, and a smaller and lighter warhead of sufficient explosive force were all reported possible, the Navy received approval to proceed with the development of the new missile. Beginning in 1957, work got underway on the Navy's Polaris missile, an IRBM launched from a submerged submarine.

The decision to proceed with Submarine Launched Ballistic Missiles (SLBM) meant that the United States would in future field a nuclear triad; that is, the deterrent force would consist of nuclear weapons delivery by manned bombers, ground-based ICBMs, and submarine-based SLBMs. This path may have been influenced by the Navy's desire to be part of the nuclear delivery ball game, and the high-priority budgets that entailed. However, this quickly fell away to a rationalization that the triad virtually assured that a retaliatory nuclear strike capability of the U.S. could not be sufficiently destroyed by a sneak attack, to prevent a devastating response on the attacker.

A submarine capable of housing and launching an IRBM did not exist; the development of the submarine would proceed in parallel with development of the missile. The initial submarine configuration was adapted from an existing attack submarine design by cutting the hull and inserting a section containing the missile launch tubes. Although several concepts with differing numbers of missiles were considered, sixteen missiles was settled on.[36] Sixteen missiles per boat was considered the best trade-off between combat flexibility and the total number of boats required to field the desired force. Such was the origin of the George Washington class of Fleet Ballistic Missile submarines. Due to this parallel development, the diameter, nominal weight, and length of the missile were established early. With this definition in place, work could proceed on the subsystems in parallel.

Due to the urgency, numerous flight tests were utilized because this could be done in lieu of waiting for complete system modeling and simulation needed to prove the concepts. Several problems were investigated and solved in this manner. The multi-stage missile required that the first stage thrust be terminated at the proper time for the second stage to deliver the payload to the proper point in space to initiate the non-powered flight path onto the target. Several test missiles were used to refine and finalize this important concept. The step-wise test program added subsystems and associated operations as the designs matured. Significant test achievements were the first flight from a surface ship, USS *Observation Island*, in August 1959; in January 1960, the first flight with the new smaller-lighter guidance system; and finally the first complete missile launch from a submerged submarine, USS *George Washington*, in July 1960. A still unusual test shot came in May 1962 when the USS *Ethan Allen* launched a Polaris A1 missile while submerged in the Pacific. The live nuclear warhead impacted precisely on target in the test range as part of the then-ongoing atomic tests. This was the only live firing of a U.S. strategic missile in history, and with subsequent treaties and testing agreements, likely will be the only such test.[37]

On November 15, 1960, the USS *George Washington* went to sea with its 16 Polaris A1 missiles for the first deployment of the new SLBM missiles. These A1 missiles had a single 200-kiloton nuclear warhead and a 1200-nm range. This was termed the initial or "interim" capability for the submarine-launched ballistic missile.

With the deployment of the initial Polaris submarines, the nation's initial hardware implementing the nuclear triad of deterrence was complete. The Polaris A1 missiles were replaced by the A2 and then the A3 missiles. These evolutionary developments of the Polaris had increasingly longer range, and the A3 carried three nuclear warheads. These longer-range submarine-launched missiles created more "sea room" for the boat, meaning that the missiles launched at a given target could be located further at sea. This greater sea room made it much more difficult for an adversary to locate the launching platform.

Between 1969 and 1976, thirty-one of the forty-one Polaris submarines were refitted with launch tubes to carry the larger Poseidon missile. The Poseidon provided a similar range capability to the Polaris A3, but could carry fourteen Multiple Independently targetable Reentry Vehicles (MIRV) per missile. Similarly, between 1982 and 1987, the Poseidon missiles were replaced with the longer-range Trident I missiles. The Trident I missile was sized to fit in the same launch tube as the Poseidon, but had a range of 6000+ nm, providing even more sea room to locate these missiles each carrying up to ten MIRV warheads.

POLARIS

The Polaris A1 and A2 missiles were two-stage, solid-fueled, inertially guided missiles carrying a single 200-kiloton nuclear warhead. The A1 missile had a range of 1000 nm while the A2 version had a range of 1500 nm. The A3 version had a range of 2500 nm and had dimensions of: length, 31 feet–6 inches; 54-inch diameter; weight 35,000 lbs. The A3 carried three 200-kiloton nuclear weapons.

POSEIDON

Two-stage, solid-fueled, inertially guided missile carrying fourteen 50-KT independently targeted reentry vehicles.
Length 34 feet, 74-inch diameter, weight 65,000 lbs.
Range: 2485 nm

TRIDENT I

Three-stage, solid-fueled, inertially guided missile carrying ten 100-KT (Mk 4) independently targeted reentry vehicles. This missile had a length and diameter so that it could be placed in the Poseidon launch tube as well as the Trident boat launcher.
Length 34 feet, 74-inch diameter, weight 70,000 lbs.
Range: 6000+ nm

The Polaris missiles were first deployed on the George Washington class of submarines. There were ten of these boats followed by thirty-one of the Lafayette class of ballistic missile submarine. The George Washington class of submarines were a modification of an existing design, with an inserted section to hold the vertical missile launch tubes. The Lafayette class was a new design to incorporate the ballistic missile operation from the start. Together these forty-one ships were called 41 for Freedom. The USS *George Washington* became operational with 16 Polaris A-1 missiles in November 1960. These forty-one boats became operational between 1960 and 1967, averaging about one new boat every four months.

As indicated above, the Polaris missiles did not have the range of the ICBMs. In order to target these missiles on the Soviet Union, the boats had to remain in the North Atlantic Ocean to remain within range. This meant that the first Polaris-equipped submarines were deployed to bases at Holy Loch, Scotland, and Rota, Spain, from where they would begin their two- to three-month-long strategic deterrent patrols. As the Lafayette class of submarines were upgraded to Poseidon missiles, beginning in 1968, the deterrent patrols began to take place farther from the Soviet Union, as the longer-range missiles would allow. Similarly, these same Lafayette class boats were further upgraded beginning in 1976 to carry the newer Trident I missiles capable of 6000+ nm range.

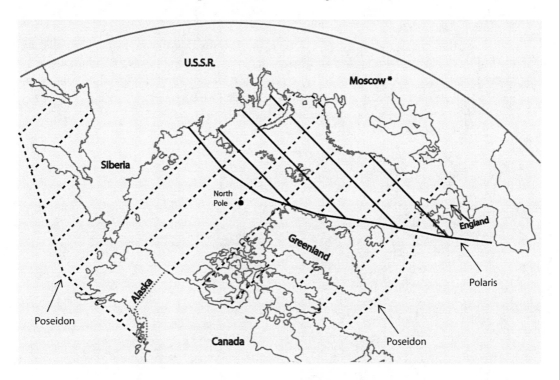

Map showing ocean areas for submarine deployment with Polaris and Poseidon SLBMs.

The Minutmen: Improved Continental Range Missiles

While the initial development, testing, and deployment of both the initial ICBMs and the IRBMs were underway, and in the postwar cost-conscious 1950s, work was also underway on the likely replacement for all the previous liquid-fueled land-based missiles: the solid-propellant Minuteman. What became the Minuteman was studied between 1955 and 1957, at which time a development program was begun. From 1958 to 1961 this project faced most of what the liquid-fueled projects had faced: cycles of moving ahead because of the possibilities, and slowdowns primarily due to the still present inter-service rivalries. These rivalries included proposals for different basing concepts, particularly mobile missiles mounted on rail cars. In early 1961 a highly successful test flight removed all but fiscal objections and production toward deployment was initiated.[38] The first 10-missile flight became operational at the time of the Cuban Missile Crisis. The Minuteman represented a simpler, safer, and cheaper system overall, and its availability allowed the Air Force to plan an earlier than expected retirement of the liquid-fueled forerunners. The deployment of the Minuteman, which would grow to 1000 missiles, overlapped considerably the completion of installations for the previous missiles. The Minuteman was deployed 50 missiles to the squadron and three squadrons to the wing. Each squadron was comprised of five flights of 10 missiles each. Each of these flights included one launch control center to monitor and launch the ten missiles. Each missile was in an underground silo and the launch control room was also in an underground protected facility. These individual sites were located no closer than three miles from one another, again to enhance survival of the force in the event of an attack.

The Minuteman is a solid-fuel, inertially guided intercontinental ballistic missile. A total of 1000 missiles were spread across several bases in Wyoming, South Dakota, North Dakota, Missouri, and Montana. A launch complex served ten of the squadron's missiles and had direct communications with those silos. The Minuteman had a range of 6300+ miles and delivered a single 1.2-megaton nuclear warhead.

Launch crew of 2:

Missile combat crew commander
Deputy missile combat crew commander

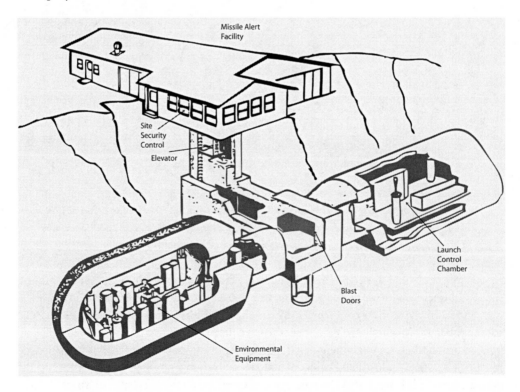

Depiction of a Minuteman Launch Control Facility.

Six bases in five states were chosen as the sites for the Minuteman missiles. Over the next several years the Minuteman replaced all the previous ground-based missiles except the Titan IIs. Of the six bases, Malmstrom AFB in Montana and F.E. Warren AFB in Wyoming were home to four squadrons each, while Ellsworth AFB in South Dakota, Minot AFB and Grand Forks AFB in North Dakota, and Whiteman AFB in Missouri were home to three squadrons each. This totaled twenty squadrons of fifty missiles each for 1000 total missiles. These installations were focused in the upper Midwest states, which allowed the 6300-mile range Minuteman missile to reach any target in the Soviet Union. Additionally, this siting arrangement placed these weapons essentially as far as possible from either continental coastline, reducing their vulnerability from aircraft or naval attack.

The following map shows the locations of the U.S. missile squadrons active at the start of the Cuban Missile Crisis in the fall of 1962. Each missile squadron contained 1, 3, 5, or 9 individual launch centers, each controlling the launch of 1, 3, 5, 9, 10, or 12 missiles.

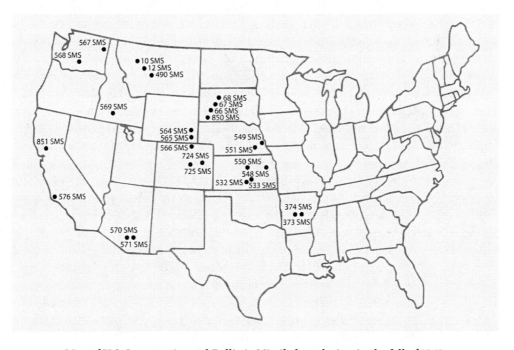

Map of U.S. Intercontinental Ballistic Missile launch sites in the fall of 1962.

At the time of the Cuban Missile Crisis, eight of the U.S. Navy's George Washington class missile boats were at sea.

MISSILE FACILITIES

By the time the Minuteman missiles were to be deployed, there were the experiences of Atlas, Thor, Jupiter, and Titan I as well as additional studies, simulations, and budgeteering to arrive at the configuration to be fielded. Ten missiles per launch facility or flight, fifty missiles per squadron, three squadrons per wing was settled on. The Launch Control Facility described below thus controlled the ten missiles of a flight.

The Missile Alert Facility (MAF) building was contained within a security fenced-in area, something like 300 feet by 300 feet. Within that area there was the main building of approximately 2000 square feet and a garage-like building that was not attached. The buildings were painted in earth tones, so that from some distance away you would see the site, but its significance would not be readily apparent; as the saying goes, the facility was hiding in plain sight. The lighting at the entry point gate, if on, would distinguish the site from, say, a farm.

The eight-person crew of the MAF, "upstairs," was on a three-day on-duty cycle. They generally wouldn't leave the facility. But of the security crew, total of 6, two might leave to investigate something at a launch site. The six security people were divided into two twelve-hour shifts of a single person manning the Site Security Control entry point and telephones, and controlling access to the underground Launch Control Facility (LCF), and a two-person security dispatch team.

In the upstairs of the building at ground level, there were two bunk rooms, one for each shift, or flight, of security people. These two rooms had cipher locks for security of their weapons. There were individual rooms for the site manager, where he had a desk and some file storage space, and a room for the cook, which was a little larger. Additionally there were two guest rooms with three or four bunks each for visitors to the site. These were not normally used by the regular crew, but were available for maintenance crews who had to stay over, or possibly for supervisory people who were visiting the sites. These extra rooms came into play later in the time period when women began to be assigned to the sites. The site manager would maintain the site, including doing small repairs himself. For other work he would schedule a crew from the base. This above-ground crew was on duty for three days at that site.

The crew would order their meals by number, which the cook would remove from the three freezers and prepare. The menu included prices so that those people on separate rations would know what they were to pay. Separate rations was the Air Force term for a person receiving compensation for his meal allocation. This was typical for married people; otherwise an individual had a meal card that allowed meals at a base dining facility. Typical prices for individual items were from .40 to .75 so that a complete meal might be 2 or 3 dollars. The cook would prepare up to four meals a day.

For recreation there was an indoor pool table, foosball table, ping pong table, and outdoor basketball hoop and volleyball court. In later times they had PC and video machines and VHS recorders. A lot of the people who worked there were working toward degrees at a state college, and others just read a lot. It was said that 1000 books were found in the upstairs rooms of an MAF when it ceased to be an active site.

The two-person crew of the LCF, "downstairs," was on a twenty-four-hour on-duty cycle. They couldn't leave the facility under any circumstances, except upon being relieved by the next crew. This two-man crew was made up of the commander and the deputy.

The security people were not allowed in the underground area, and there would be no reason for the launch crew to be in the above-ground spaces except during entry and exit. Only the cook was allowed to enter the elevator to take food down to the launch crew. However, this was difficult since the launch crew would have to retract the pins from the blast door to allow the cook to pass food through the door, and they would then have to re-secure the blast door, a time-consuming and somewhat physical process. As a result, the launch crew normally carried their own food for their 24-hour shift.

The upstairs security facility was a Site Security Control (SSC) that contained telephones to communicate any required notification to the command post, and radio to communicate with mobile units. The individual working there was the senior person of the security crew on duty, and was called the flight security controller.

The windows of the security control point were covered so that someone outside the building could not see where the individual was. He worked in the room blacked out except for backlit panels that contained key procedure checklists and key information such as phone numbers. These procedures and checklists were under glass so that he could write on them in wax pencil.

The MAF had no missile at the site. The nearest missile would be 3 miles or more away, and the spacing between missiles would be 3 miles or more. This created a more difficult targeting problem for the Soviet Union, so that each site would require a missile to take it out. Each of the halves of the underground portion of the Launch Control Facility (LCF) were a cylindrical steel room with the floor suspended on isolators that were to preserve the building from a near-miss attack. The building would not survive a direct hit, but this preparation was to enable it to survive the near miss. Between the floor and the walls of the cylinder there was room for the floor to move. Construction of this facility entailed first burying this steel cylinder and building around it a structure of reinforcing steel rebar, which was filled with concrete, all of this 31 feet underground. At completion the wall around these cylinders was approximately 4 feet thick.

The non-launch room chamber contained mostly environmental equipment including an air filtering system to filter the intake to the facility such that it could remove chemical, biological, and radioactive contaminants from the air to have continuous clean air for the launch crew. The room also contained a diesel generator to provide backup power to the normal commercial power supplied to the facility. There was fuel on board for about 9 days, and there was also food stored under the floor for about 9 days. There was an escape tunnel behind and toward the end of the support chamber. This tube, which went from the chamber to the surface, was normally filled with sand, which would fall into the chamber when the crew removed the plug. The crew would have to dig out manually the last five feet of the tube's length up to the surface to escape. No one knew where this tube came to the surface; it was not marked. This was an escape path for the crew, presumably after an attack on the site.

In order to enter the LCF control chamber, the arriving crew would check into the site at the Site Security Control point upstairs and proceed through the security control room into the elevator. Access to this elevator was first controlled by the flight security controller and verified by voice communication with the on-duty launch crew by exchange of a number code. This elevator was the only method of moving from the upstairs to the underground chambers. The crew on duty would have to unlock the launch chamber, which entailed removing the locking bolts from the chamber blast door. This was an entirely manual process by which a crewman would operate a hand hydraulic pump, several hundred strokes, to first retract the bolts, allowing the blast door to open. Near completion of the changeover, the departing crew would remove their personal locks from the secure lock box, and the arriving crew would install their locks. Then following the changeover procedure, the new crew would operate the pump to re-engage the locking bolts of the blast door. This operation took minutes, and made it practical for the launch crew to remain in the locked chamber for their entire twenty-four hour shift.

The launch chamber has two operator consoles, numerous electronic equipment racks, a single bunk where one of the individuals could sleep, a small food preparation area, and a small toilet area. The walkway was essentially L-shaped and allowed for moving around, but they could not go very far. If one of the launch controllers got sick, it was theoretically possible for a replacement to come out from the base; it was said this never happened.

The consoles at the LCF had two sets of status lights for the ten missiles of their flight. These lights were connected to each missile site and lights indicated the status of the missile, the faults, and the security alarm at the site. The LCF is only directly wired to the missiles of that flight. Each desk had a phone that could connect directly to the phone at each launch site with just a button push. The launch officer could call the phone at the launchsite, to speak to either security or maintenance people, if present. Additionally, he could also talk to the other flight launch facilities, at least four sites. He could also make a normal land line phone call such as to communicate with his squadron offices or the security office upstairs. There was a terminal at the commander's station in the launch chamber that allowed him to re-target an individual missile.

There were four communications paths for the launch order to be delivered to the LCF launch crew. The Primary Alerting System communication was by voice land line. There were two teletype systems at the deputy's station as well as radio. A UHF antenna was outside to receive communications from a Looking Glass airborne command post aircraft. A warbling tone would alert the crew that an emergency war order was arriving and for them to copy the information onto a checklist. Following receipt of the message, each crew member would unlock his individual locks, total of two, on the secure lock box containing the launch keys and the authenticating "cookies." The cookies were plastic cases about the size of cookies which the crew would break to release the information each needed to authenticate the received message. If the message was authenticated by both officers, they would then commence the launch sequence, culminating in inserting their individual launch keys in the rotating switches and turning the keys within a short time frame. The location of the launch key switches was twelve feet apart by design so that it was impossible for a single individual to insert and turn both keys in the small allowed time span.

The received message included the launch code, which the deputy inserted into thumb wheel switches on his console. This launch code contained the actual command to the missiles of the flight. This code might launch all or a portion of the flight's missiles; the crew did not know. The crew also did not know the target location for the missiles. The two-man policy in place at each launch control facility required both the commander and the deputy to agree that the launch order was valid, and then insert and turn their respective keys, in order to actually launch missiles.

The two missile launch officers, or missileers, would monitor the signals from the launch sites. The signals included the security alarm of the remote sites. This seismic alarm was routed to the monitoring panel for the missiles monitored by the launch crew. If an alarm occurred, the launch officer would have to telephone to the site security controller upstairs, who would initiate the response by the security force.

Many of the MAF, LCF, and launch silos were in use for 30 years or more, from the early to mid–1960s to the mid or late 1990s. In the 1980s these buildings were refurbished to make a little more recreation space. The dining room was remade to have individual tables with pictures on the walls. Of note was the paint scheme, which was changed from the earlier dark green to more earth tones. When the buildings were first built, they were not set up for women, and so the early period of having women assigned just meant a sign on the bathroom door that could be flipped. Upgrades to the LCFs included taking part of a storage room and converting it to a women's bathroom. In the launch chamber, where previously there had been a curtain as partition for the toilet space, in the later scheme this was replaced by a small door for privacy.

LAUNCH SITE (SILO)

There were signs on the highways near intersections to direct personnel to exit the highway at a particular intersection. These signs were large enough to be seen from the roadway, but their meaning was not immediately obvious to others. Such a sign would just have "B-4" in white letters on a brown background.

Each launch site was a small fenced-in area approximately 150 feet square. The entrance gate would be perhaps 30 feet from the secondary road near the site. This secondary road could be gravel and was most often a county road. A launch site, or silo, was unmanned, and Air Force personnel only visited when needed. The fence included a locked gate, the entrances to the underground were secured by a series of locks, and sensors at the site were monitored at the Launch Control Facility, several miles away.

The underground missile silo was covered by a 90-ton cover, mounted on rails, so it could be moved to open the top of the silo. Prior to missile launch, the cover would be moved by an explosive charge, which would leave the cover several yards from the silo.

Also visible above ground at the launch facility were the soft support building, the personnel access hatch, the physical security system, the hardened UHF antenna, and two pylons near the silo.

The soft support building was made of reinforced concrete, of which about one foot protruded above ground. The underground building housed a backup diesel generator and a chiller unit to provide controlled temperature and humidity for the missile components. This controlled environment was crucial, particularly for the electronics. The site was on commercial power as the primary, but for any failure, the generator would kick in to maintain the site at complete readiness. The temperature inside the silo was maintained very near 60 degrees.

To access the underground silo, primarily for maintenance work, security personnel would open a lid adjacent to the hatch, and enter the code of a combination lock. This would cause the hatch, a heavily reinforced steel and concrete door, to begin its several-minutes-long opening. Through this open hatch the maintenance crew would access another combination lock. When this combination was successfully entered, a steel door called the V plug moved to the level of the equipment room inside and access was gained. The interior spaces of the underground areas were "no lone zones," that is, any individual must be accompanied by another person of similar knowledge level and they must maintain visual contact always.

The physical security system of the site was to detect intrusion into the site and motion along the perimeter fence. A detection from the security system would cause an alarm at the panel in the LCF. Such an alarm would cause the launch crew to notify the flight security controller of the alarm. The security controller would then dispatch the two-man security team from the LCF to investigate. Most often such a detection would be caused by either an animal such as a deer or rabbit, or caused by the farmer on the nearby field operating machinery. It was said that although the land owner was well compensated for the government's use of this land by a long-term arrangement, this did not prevent the farmer from planting right up to the fence boundary and hence extracting all possible economic benefit. But this practice did contribute to additional security alarms over the years.

The hardened ultra-high frequency (UHF) antenna was a radio antenna protected by concrete against nuclear attack. The antenna provided a communication path through which the SAC Airborne Command Post, known as Looking Glass, could transmit a launch command directly to this underground silo. This equipment provided a backup plan in the event that the SAC Command Post at Offutt AFB was damaged or the flight LCF was damaged, or both.

Two pylons located near the silo cover were used to align the missile transporter trailer for installing the missile into the silo. A specialty vehicle called the Missile Transporter Erector moved the missile from the main base to the silo site. Once at the site, this vehicle would be backed into position where the trailer could be set down onto the pylons. The trailer would then be hoisted into the vertical position, completely covering the silo. From this position the missile could slowly be lowered into the silo. Placing the missile into the silo normally took 3 to 4 hours. Next, the warhead would be brought from the base separately and installed atop the missile.

After this missile had been placed inside the silo, the missile had to be aligned to fix it precisely in the coordinate system of the navigation calculations. With a 6000+ mile flight path, even the smallest initial error could cause the impact location to be off by several miles. To aid in this alignment, near the bottom of the Minuteman I silo there was a ledge that mounted an angular scale that ran completely around the inside of the silo. This scale had been very carefully installed and shifted to align to north at that specific spot on the earth, following construction of the silo. The alignment process required mounting an optical instrument on the angular scale ledge, with which angular measurements of stars were made. Measurements from three stars were completed over three days' time to produce the necessary information to fix the missile's position. Other methods of fixing the precise location of the missile for later versions of the Minuteman were adopted.

Missile Flight

Missiles are divided into two class definitions: guided and ballistic. A missile is considered guided if there is a mechanism present to adjust the flight path for most of the flight time. A missile is considered ballistic if there is no mechanism affecting the flight path for most of the flight time.

The long-range missiles discussed here are ballistic in that the powered rocket boost phase immediately after launch is controlled by a guidance scheme to put it and keep it on its intended path. But this powered and guided portion is a small fraction of the total flight time, perhaps 100 or 200 seconds out of the total flight time of 20 to 30 minutes.

The flight path of a ballistic missile is divided into three distinct phases: the powered and guided phase, the free flight outside the atmosphere, and the reentry or terminal phase. These phases are distinguished by the forces acting on the missile during the phase. Only in the powered phase are the rocket thrust forces present. Only in the exo-atmospheric free-flight phase are no atmospheric, i.e., aerodynamic, forces present. The force of gravity is present in all phases.[39]

It is practical to determine the flight path of a missile by working from the desired target back to the launch point. For example, for the payload to land at a certain target location, it must have reentered the atmosphere at a certain range to the target at a certain speed; to reach this point at the conclusion of the free-flight portion of the flight, it must have entered the free-flight portion at a specific altitude, speed, and angle at the conclusion of the powered and guided portion of the flight; and to reach this point at burnout of the rocket, the guidance system had to steer the missile to this angle at the speed attained.

The resulting path is very nearly a portion of an ellipse with one of the foci located at the center of the earth. The exact shape of the ellipse, its eccentricity, determines the range of the path. It is the job of the missile's guidance system to steer the powered missile onto the specific elliptical path that will bring the payload to the precise target location following the uncontrolled free-flight and reentry portions of flight.

For weapon delivery at this range of nearly 10,000 miles, the ballistic missile flight path has advantages over the other possible forms of delivery. By launching the missile in the vertical direction and then turning to the desired angle just before burnout, the structure of the missile is simplified and of lighter weight because the launch vehicle does not have to withstand side loads resulting from aerodynamic affects. This is important because extra weight means less range. For example, on one U.S. missile, one pound of additional weight at launch reduced the maximum range by one mile. As a result of the payload's being propelled high above the earth fol-

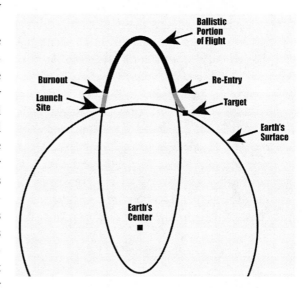

Diagram of the flight path of a ballistic missile.

lowing launch, up to 750 miles, two advantages are attained: first, the force of gravity accelerates the falling payload to a very high terminal speed, around 10,000 miles per hour, allowing a very small time for the adversary to intercept and interfere with the flight. And second, the high position above earth, combined with the straight line of the free-flight portion of the path, results in the payload's landing in a near vertical path, resulting in a smaller dispersion about the target location, compared to the other forms of delivery. Because the majority of the flight path of such a missile is outside the atmosphere, and hence outside the effects of the atmosphere on the flight, the amount of energy needed, i.e., fuel, to achieve the same range of attack is smaller, again compared to the other forms of delivery.

Perhaps the drawback to such a missile is that since the powered and guided phase is short compared to the total flight time, essentially the target location cannot change following launch, making only a predetermined target practical for these weapons. Another requirement of the payload that reenters the atmosphere is that the combination of the great speed and increasing air density as the package descends, causing aerodynamic drag, is that the package must be designed to withstand the tremendous heat generated during the final descent. A typical ballistic missile warhead package would reach over 1500 degrees C during reentry.

Another reason for the retirement of the earlier liquid-fueled missiles fairly soon after deployment was the number of crew members required for the earliest designs. The Atlas D missile launch facility required a crew of twelve, the Atlas E and Fs required five, Titan I required four, and the Titan II four. Additionally, with the Minuteman the launch crew of two served a flight of ten missiles. With each advance of the missile technology, more conditions were monitored by electronics that could alert the crew with an alarm, reducing the need for a specialist to monitor that condition.

The defense establishment learned very quickly with the early missiles that it was not required for a single crew be responsible for a single missile. The arrangement of ten missiles per launch crew was arrived at as the best mix for flexibility, reliability, and economy. This has been the arrangement for the U.S. land-based missiles from the mid–1960s through to the end of the Cold War and remains the case today.

The long-range ballistic missile was not an effective approach to attack a moving target such as a train, or a convoy of either ships or ground vehicles. This tended to preserve the need for bombing aircraft to attack either a field army or a naval force.

The force consisting of B-52 manned bombers supported by KC-135 refueling tankers, the 1000 Minuteman ground-based ICBMs, 54 Titan II ICBMs, and forty-one strategic submarines each housing up to sixteen IRBMs, was the U.S. strategic deterrent throughout most of the Cold War. This triad of delivery methods could deliver nuclear weapons on their specific targets upon order of the president of the United States. No other authority could initiate the launch of this massive destructive capability.

Chapter 5

Sky Watching

The U.S. experience in World War II had included defending against attack by cruise missiles, specifically the German V-1, but there was no history of defending against longer-range, higher-flying missiles. The cruise missile has similar flight characteristics as the airplane in terms of speed and altitude. As a result, the air defense weapons in use against airplanes could be adapted to the World War II V-1. This was not possible for the V-2 rocket; the U.S. had no defense against the longer-range, higher-altitude, and higher-speed missile of World War II. During the Cold War period, capability to monitor for a rocket-delivered attack had to evolve as the newer weapons came to the fore.

Collecting the Information at a Single Site called NORAD

The establishment of a single command and control center for the air defense of the United States had come about slowly in the years following World War II. Beginning in 1946, the Air Defense Command of the U.S. Army, and then the Air Force, had the responsibility to operate the air defense radar sensing network and the fighter interceptor aircraft to respond to detections made by the system of sensors. Admiral Arthur Radford, then Chairman of the U.S. Joint Chiefs of Staff, believed that the contributing forces to the combined national air defense should be under a single operational command arrangement. This led to the establishment of the U.S. Continental Air Defense Command (CONAD) in 1954.[1] The Army's element was the Antiaircraft Command, which operated the air defense artillery. This became the Army Air Defense Command (ARADCOM) in 1957, following its transition to the Nike family of surface-to-air missiles. The U.S. Navy provided radar-equipped picket ships along the East and West coasts of the U.S., from the Naval Forces for CONAD.[2] This arrangement led into the establishment of the North American Air Defense Command (NORAD) headquartered in Colorado Springs with a U.S. Air Force general in command.[3]

NORAD's first headquarters was located in a refurbished hospital building in Colorado Springs, Colorado. A key concept of the evolution mentioned above was the single and central nature of this command center. This called for channeling all pertinent information on an attack, and on the status of response forces, to this one building and staff of specialists for evaluation so that any needed order to change the readiness of those forces could be issued. An obvious weakness of the initial implementation was that the entire defense system could be neutralized with the destruction of this single building with its personnel. As Gen. Earle Partridge, the first commander of NORAD, told his superiors, "[A] man with a bazooka in a passing car could put the establishment out of commission."[4] In 1959 the Joint Chiefs of Staff approved a project to create an underground facility capable of withstanding nuclear attack and continuing to function as planned.

Located underneath 9565-foot Cheyenne Mountain several miles outside Colorado Springs, the NORAD Combat Operations Center (COC) was (and still is) essentially a small city inside a mountain. The eleven buildings located inside the mountain are under at least 1400 feet of rock. After entering the mountain two thousand feet along a mile-long access tunnel, people entered through a passageway that could be secured by two 30-ton blast doors on either side of a fifty-foot isolation section. These doors would be opened one at a time, to maintain protection of the interior, during a time of crisis. For additional protection from nuclear attack, all the buildings are mounted on shock-isolating coil springs, almost a thousand total. Inside this mountain are fresh water reservoirs, storage for food and other supplies, electrical generating equipment, air filtration equipment, and an underground reservoir of oil to operate the generators. Sleeping and dining facilities to house 900 people are also present in addition to supporting medical facilities. The provision is to house these 900 for over thirty days following an attack.[5]

NORAD was initially responsible for the air defense of North America, established to defend against attack by manned bomber aircraft, detection of which was primarily provided by the DEW Line radars. These responsibilities grew to include similar preparations for attack by ICBMs and submarines launching nuclear armed missiles. Monitoring for attack by missiles was the function of the Ballistic Missile Defense Center (BMDC), located within the NORAD Combat Operations Center. This section provided the link between the NORAD commander and the Safeguard defense sites, after Safeguard had been deployed. Other centers within the COC were stations to monitor objects in space, particularly in earth orbit, and to monitor reports from coastal radars that would indicate the launch of a ballistic missile from a submarine. These sections were called the Space Defense Center (SDC) and the SLBM Detection and Warning System. Improvements in radar were required to accomplish these protections.

RADAR EVOLUTION

The radar sensing installations discussed so far were based on the technology and methods developed prior to and during World War II. Substantially, a continuous string of individual pulses of transmitted radio frequency energy was directed into the region of air to be monitored. An object in this region with suitable physical makeup created a reflection of the pulse, which radiated away from the reflecting object, including in the direction of the transmitter location. The reflected signal was received and processed by the radar. Some of the earliest radars used fixed antennas to transmit and receive these signals. Before the end of World War II the more common installation was for a rotating antenna to direct a beam narrow in azimuth through a circle to cover 360° in azimuth over the duration of this rotation. This provided a return signal from each object in this region once during this rotation, also called the radar sweep. This sweep operation was augmented by a feature in the control of the antenna that would limit the sweep to a fraction of the entire 360°, say from 45° to 70°. This allowed an operator to maintain a close watch on the smaller sector, which provided more returns over time from an object in that sector.

This mechanically swept antenna was adequate as long as you were only interested in objects from, say, the 45° to 70° sector. This rotating mechanism was not good at monitoring an object at, say, 50° and another at, say, 310°; to do this the antenna would need to be swept from about 305° to 55°, perhaps better than the 360° sweep, but something else was needed.

This need led to the development of a radar that could send a beam in any desired direction as opposed to being sent only in the direction of the rotating antenna. This technique is termed the phased array radar because the antenna face does not move; the phase of the signals transmitted from different parts of the face are adjusted so that the combined signal made up of the individual signals from all parts of the face are stronger in a specific direction. This direction and its corresponding set of phase adjustments are established by conditions set for each contributing signal. If these settings were constant, the beam would be pointed in a specific direction. The later extension of the principle was to make these settings a command from a digital computer. As a result, the digital computer could command an arbitrary angle for the beam at one time, and an unrelated angle for the beam a short time later; the mechanically swept beam could only move left-right and up-down, limiting the beam to continuous movement in a region.

Following World War II and the experience of the threat of attack by German V-2 rockets, the U.S. conducted experiments in the detection of such weapons in 1948 and 1949 using existing radars. These efforts fueled the belief that much more powerful radars providing long-range detection were possible. In the early 1950s, information that the Soviet Union was developing military rockets began to emerge. However, due to the Iron Curtain established around the Soviet territory and that of its satellites, information on the progress of the developments was limited. Particularly the degree to which the U.S. was under threat of an ICBM attack was not sufficient. When it was apparent that airborne weapons delivery would evolve to the long-range missile, the sensing and response capabilities of U.S. air defense was slated to evolve further to combat the newer method of attack. As a result, the U.S. undertook the development of a long-range radar that could detect and track the Soviet test launches from outside their territory. The result was the construction of what came to be called the FPS-17 radar, built by the General Electric Company, installed at Diyarbakir, Turkey, in 1955. GE was selected in part because a key to the new radar in a short time was the use of high-power transmitter components that company had developed for the relatively new broadcast television industry. The antenna was a large parabolic reflector, 175 feet high by 110 feet wide, and was fixed in position generating a fixed beam pattern.[6]

The site in Turkey was hardly arbitrary. The radar at this site was to monitor the air above and beyond Kapustin Yar, slightly east of Volgograd, USSR (then Stalingrad), for missile launches. Kapustin Yar was one of the launch sites of the Soviet long-range rocket program. Launches from this site tended to be tests of one or two components of a complete missile. The flight path of such a launch was across Soviet territory to the southeast, ending near Lake Balgash, in the vicinity of Saryshagan. Missile flights on this range were limited to a distance of around 1100 nautical miles.

The data from a missile flight was captured by photographing the display screen of the radar with a moving film camera. This film was then analyzed to extract the sequence of positions that the missile had flown through in flight. From this data analysts derived the speed and altitude and hence extracted the performance of the missile itself.[7]

The need for the intelligence on the Soviet missile developments was so crucial that the initial installation of this new radar was in Turkey to observe the Soviet developments beginning in June 1955. A second FPS-17 installed at Laredo, Texas, became operational in February 1956, and allowed for a longer-term study of the radar itself. The final copy of this radar was installed on the island of Shemya in the Aleutian Islands of Alaska, where it became operational in May 1960. The Shemya site, nearly at the westernmost end of the Aleutian chain, was ideal to observe the terminal end of missile flights that originated at the Tyura Tam launch site.[8] Tyura Tam was located in Kazakhstan slightly east of the Aral Sea. The flight path of such an ICBM launch was generally northeast toward the Kamchatka Peninsula on the Pacific Ocean side of the country. Flights on this range could reach a distance of around 3500 nautical miles. Some of the reentry packages of these tests landed on the Kamchatka Peninsula and others landed in the waters of the Pacific further east. The launching complex at Tyura Tam would later be known as the Baikonur Cosmodrome, ever since a center of Soviet space activities.

A second type of radar was needed to track a missile once detected to accurately predict the landing area so that the correct defense site could be alerted to the attack. This project was assigned to the Lincoln Laboratory of MIT. For this assignment Lincoln constructed a prototype radar at Millstone Hill in Westfield, Massachusetts. To test the new concepts involved in this radar, missile launches from Cape Canaveral, Florida, later renamed Cape Kennedy, were monitored and tracked. These new concepts were Doppler processing, higher-power components, and computer tracking of targets. All these concepts needed to be proven reliable for the long-range missile detection and tracking system to work. The Millstone Hill radar consisted of an 84-foot parabolic reflector mounted on an 85-foot pedestal base. The associated digital computer controlled the azimuth and elevation angles of the antenna's position in addition to processing the radar signal returns.[9]

ANTI-MISSILE MISSILES

While these developments in the radar needed to provide early warning of a missile attack were underway, events were also underway to provide for the weapons to oppose such a missile attack. Prior to this the Army had been developing the Nike air defense missiles to bring down attacking bomber aircraft. As these were in their initial deployments for the bomber threat, the Army began the research work to expand these concepts to counter an attacking missile. This was known as Nike Zeus. The Air Force had a study project for an anti-ballistic missile (ABM) system as well, called Project Wizard. This project included the long-range detection of an enemy launch and the defensive missile that would be launched to defeat the enemy missile.

These projects were another example of the inter-service rivalry between the Army and Air Force over missile weapons. During this period of the mid–1950s, both services had active developments underway for intermediate-range ballistic missiles (Jupiter and Thor), surface-to-air missiles (Nike and BOMARC), and now anti-ballistic missiles (Nike and Wizard). In 1956, Secretary of Defense Charles Wilson instituted a change to these contested areas of the roles and missions of the service branches. He directed that the Army would be responsible for point defense, meaning surface weapons out to a 200-mile range and air defense out to a 100-mile distance. The Air Force would be responsible for area defense beyond those ranges. This meant that the Jupiter missile was transferred to the Air

Force for deployment. Following the successful flight of the first Soviet ICBM in August 1957, the next Secretary of Defense, Neil McElroy, further refined the mission assignments of the two services. The Army was assigned the task of developing the anti-missile weapon, the Air Force Wizard ABM work was canceled, and the Air Force was directed to focus on the development of the long-range radars needed for the defense against missile attack.[10]

Ballistic Missile Early Warning System

In 1958, at the time the Air Force was directed to provide for the long-range early warning radar detection of an approaching ICBM-delivered warhead, work was underway on the radar elements of the Ballistic Missile Early Warning System (BMEWS). NORAD was directed that it would be the central command and control agency for handling of missile warnings generated by the early warning system. This new task assignment was further incentive to secure the command and control elements of the early warning for the nation's defense. The new NORAD command post was to be established in a single, central, and hardened facility near Colorado Springs.

BMEWS was created to monitor for missile approach across the Northern Hemisphere launched from the Soviet Union. Begun in 1958, the station in Thule, Greenland, which monitored most of the Soviet heartland, became operational in 1960. The nature of the ICBM flight required a different approach to provide detection; an ICBM path would completely bypass previous radar lines, which covered from near the surface to, say, 100,000 feet altitude. The short (approximately 30-minute) flight time required further automation to alert probable target areas. To obtain the longest possible warning time, as with the aircraft detection radars, the sensors were placed on territory controlled by the U.S. or its allies as close to the borders of the Soviet Union as possible.

Additional sites at Clear, Alaska, and Fylingdales Moor, Yorkshire, England, completed the sensor network. The U.K. station did double duty in that it could also detect shorter-range missiles launched from Eastern or Central Europe toward Western Europe or England.

The success and concepts of the FPS-17 radar were extended into the design of the FPS-50 radar for the long-range detection element of BMEWS. General Electric built a prototype of this system on the island of Trinidad in the Caribbean, where in addition to initial testing, it monitored U.S. missile tests over the Atlantic Missile Range and launches from Cape Canaveral. The experimental radar at Millstone Hill served as the prototype for RCA's FPS-49, which became the pencil-beam tracking element of the BMEWS.[11]

The BMEWS radar detection concept called for three fixed orientation antenna faces at each site to transmit sets of two beams, a low beam and a high beam. The low beam was at 3.5° elevation and the high beam was at 7° elevation. Each individual antenna face produced a beam 40° wide; together the three antennas generated a total beam of 120° in the horizontal. Pairs of these fan-shaped beams would be directed into the sky to cover the region. Detection of a missile would occur when an object first produced a return from the low beam, followed by a return from its mated high beam. These two returns in proper succession provided the computer with the raw data to predict the launch area and the target impact area for the ballistic missile. This was the operation performed by the FPS-50 radar.

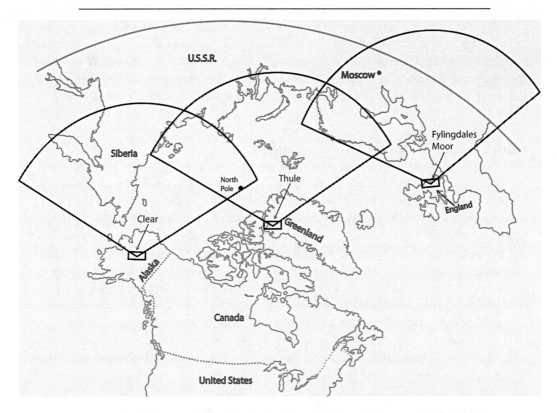

The installations of the Ballistic Missile Early Warning System (BMEWS).

The BMEWS system included a separate radar for tracking the detected ICBM. The 84-foot tracking radar antenna was housed inside a protective radome 140 feet in diameter. It could track an object the size of a house door at 1500 nautical miles distant. This was the operation performed by the FPS-49 radar. A detection by one of these radars was communicated to the NORAD center in Colorado Springs through extensions to the DEW Line rearward communication system, previously described in the section on the DEW Line radars. The combination of the FPS-50 and FPS-49 became operational for the BMEWS in 1960 and served through most of the Cold War. In 1987 these radars began to be replaced by new radars very similar to the Pave Paws equipment discussed later in this chapter.[12]

Anti-Missile Missiles: Nike Zeus or X

Even before the air defenses against manned bombers described in a previous chapter were complete, work began in earnest to provide for a defense from attack by intercontinental ballistic missiles carrying nuclear bombs. In response, the Army, the developer and operator of the Nike air defense system, initiated studies to evolve an anti-ballistic missile (ABM) capability out of Nike to protect against ICBM attacks. Initial studies begun in 1955 led to Nike Zeus development by 1958.

The initial efforts to establish anti-ballistic missile (ABM) defense was called Nike Zeus. Although quite similar to the fielded Nike Hercules in concept, this system required more powerful radars and a more capable missile, to achieve intercept of the much higher speed target. The concept of Nike Zeus was that a long-range radar would detect the Soviet missile launch, determine the probable target area, and report this information to the command and control center. Based on the probable target area, the information would be passed to a defense battery, and specifically to the Target Track Radar of that battery. The target tracking would continuously update the specifics of the inbound target and feed information to the computer, which determined the precise instant to launch the interceptor missile. After launch of the interceptor missile, the Missile Track Radar (MTR) would guide the interceptor to the inbound missile. Through the MTR, the computer would send a command to the missile to detonate its nuclear warhead at a point close enough to the inbound missile to destroy it.[13]

ZEUS SYSTEM COMPONENTS

The Nike Zeus system went through three stages of evolution prior to eventual deployment of a system. Much of this evolution was in the radar hardware as various challenges were identified and overcome. Initially there was a Forward Acquisition Radar (FAR), a Local Acquisition Radar (LAR), and the Target Track Radar (TTR). This set certainly showed its heritage from the Hercules equipment. These radars were all of the rotating antenna variety. The next revision combined the FAR and LAR into the ZAR for Zeus Acquisition Radar. The ZAR was to have 1000-mile detection range. The Decoy Discrimination Radar (DDR) was added at this stage. A decoy was a device released from the enemy missile along with the nuclear warhead, or warheads, and was intended to confuse the defense system. The general concept was that for each warhead there could be one or more decoys for the defense to sort out. At this point in the evolution of the radar hardware, an additional radar was needed to distinguish the actual targets from the decoys. The following step in the evolution was to the name set of the Nike X: the Multi-function Array Radar (MAR) and the Missile Site Radar (MSR).

Testing of these system components was accomplished via test installations at White Sands Missile Range (WSMR), in southern New Mexico. The complete prototype Zeus system was tested by installation of the individual radars installed at the Kwajalein Atoll in the central Pacific Ocean. In late 1962 the prototype Zeus system successfully intercepted a U.S. Atlas missile over the Pacific. Additionally, successful tests included the interception of a Hercules missile by the Zeus system, on two separate tests. However, testing of the prototype Zeus system with continued study and evaluation of possible attack scenarios, between 1958 and 1963, led to the conclusion that the Nike Zeus system should not be deployed. Rather, work should continue on more advanced concepts, particularly for the radars, missiles, and computers. The inability of the Zeus system to distinguish between reentering objects, and the likelihood of being overloaded by the sheer number of reentering targets, were major factors in the decision to not deploy that generation of hardware and system. This more advanced set of concepts became known as Nike X by 1963.

The Nike X concept replaced the older mechanically steered rotating radar antennas with the newer technology of electronically steered antenna beams, known as phased array radar. Central to the concept was the Multi-function Array Radar, or MAR. This radar would acquire and track the inbound target from data provided by BMEWS. The ongoing tracking would enable the MAR's computer to predict the precise expected future position and so assign that target to a particular Missile Site Radar, or MSR. As the name implies, the MSR provided guidance for the missiles of a particular site installation. For an area defense, there could be multiple MSRs and their associated missile sites spread around the perimeter of the area being defended, such as an urban population center. The several tasks of the MAR were to detect, track, discriminate, and guide a weapon to intercept a particular target.

The weapons of the Nike X system were an improved Zeus missile of 400-mile range, intended to intercept and destroy the target hundreds of miles up and hundreds of miles away from American cities, and the new element, the Sprint missile. The higher-performance version of the Zeus missile was named Spartan. The Sprint missile was extremely high speed and was intended for use when a target had nearly reached the edge of the atmosphere, and where the Zeus missile would not be effective. This situation might occur, for instance, when it took longer for the computer to discriminate the actual warhead from other objects, leaving insufficient time for the Spartan missile to reach the target.

The MAR of the Nike X system, primarily through its computer, needed to discriminate the actual nuclear warhead from other objects carried by the missile. The actual warhead or warheads needed to be differentiated from dummy objects, known in engineer-ese as penetration aids, and other possible debris from the launch rocket. The computer could most easily distinguish between a warhead and other objects when the object got very close to the earth's atmosphere and the motion of the object began to be influenced by atmospheric effects. This discrimination by the MAR overcame one of the principal objections to Zeus: the handling required of multiple objects ejected from a missile.[14]

The other major impediment that held up the commitment to spend significantly to deploy a missile defense system was the notion that, once deployed, the Soviets could devise ways to counter the system's impact on their offensive capability. The ABM could be countered by additional Soviet missiles and warheads. These ideas are combined in the concept that the additional offensive weapons necessary to overcome a level of defense installation are less expensive than the next level of defense capability. This basic advantage for offensive weapons creates the specter of a continuing cycle of response to an opponent's improvements, commonly known as an arms race. It was particularly Robert McNamara, Secretary of Defense from 1961 to 1968, who advanced this concept, and as a result did not approve spending for deployment of a missile defense system during most of his tenure. McNamara saw a further arms race as unproductive for both sides in that the balance of destructiveness, or terror if you will, could always be a step away from imbalance. That potential for imbalance would be attractive for one side's taking advantage of the opponent's weakness. He proposed talks with the Soviets to limit defensive weapons. Such limiting would in effect preserve the ability of both sides to thoroughly destroy the other, the concept of Mutual Assured Destruction (MAD).[15]

Both Nike Zeus and Nike X were proposed for deployment several times between 1961 and 1966. The Army argued for the value of establishing even a limited capability while the system continued to mature. Opponents argued principally that such a system was inherently destabilizing and its deployment invited a first-strike attack on America.

Deploying an ABM: Sentinel and Safeguard

With little or no progress toward a negotiated limitation, the Nike X system was approved for deployment as the Sentinel Defense System in 1967 during the administration of President Lyndon Johnson.[16] The renamed Nike X system was planned for deployment around U.S. cities following emergence of a Chinese nuclear missile threat. Seventeen population centers were scheduled to receive these defensive installations. However, with the administration of President Richard Nixon taking over the reins of government in early 1969, a complete review of the controversial deployment decision was conducted. Nixon replaced Sentinel, oriented to protect population centers, with what became known as Safeguard to protect ICBM missile fields, and other sites associated with U.S. retaliatory capabilities. Under this plan, twelve sites were scheduled for installations of defensive radars, computer centers to accomplish the data processing, and their associated underground missiles.[17]

The Safeguard system was comprised of the perimeter acquisition radar (PAR), which was an upgraded MAR of the Nike X parlance; the Missile Site Radar (MSR) retaining the

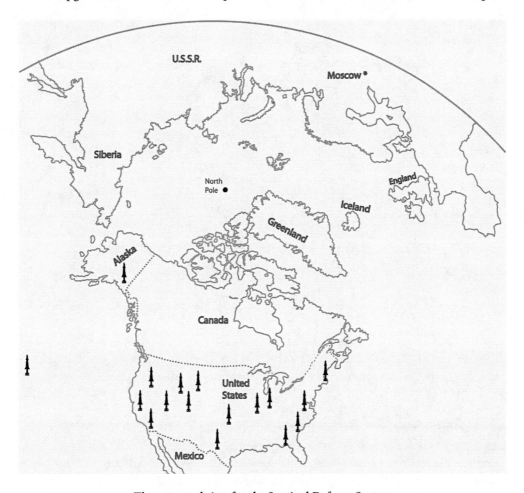

The proposed sites for the Sentinel Defense System.

name from Nike X; Spartan missiles; and the shorter-range Sprint missiles. Both the Spartan and Sprint missiles were located in the common missile farm complex adjacent to the MSR, and were both guided following launch by the MSR and its computer. The Safeguard system included the capability for the MSR to control additional Sprint missiles located within 25 miles of the MSR site.

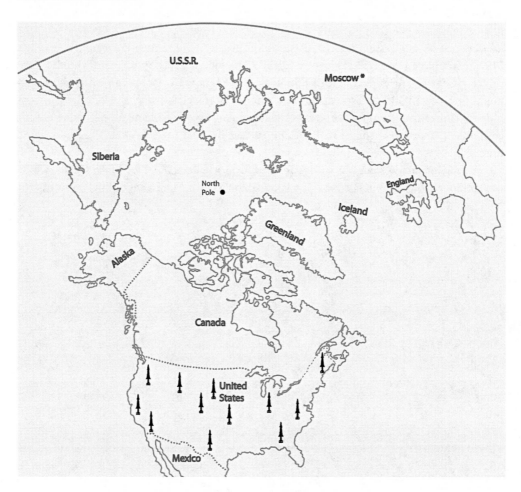

The proposed sites for the Safeguard Defense System.

Each of these capabilities of the MAR were tested at the Kwajalein Atoll test range. Navy Polaris IRBMs and Air Force Minuteman ICBMs were both used as targets in the live tests. These tests were monitored by Soviet ships offshore of the Kwajalein Islands. Monitoring the long set of system tests undoubtedly provided Soviet leaders with a complete and accurate picture of the ABM system's capabilities.

The first Safeguard PAR was constructed at Cavalier, North Dakota. The first MSR and adjacent missile farm was at nearby Nekoma. In this site, named the Stanley R. Mickelsen Safeguard Complex, the missile farm contained underground launchers for thirty Spartan and sixteen Sprint missiles. In the North Dakota installation, four remote sites were outfitted with Sprint missiles controlled by the central MSR.

The 1972 ABM Treaty with the Soviet Union limited each country to two deployed systems with geographic and numerical limitations: one to protect the seat of government, the national command authority, and a second to protect some missile launch sites. As a result, the Soviets retained their system to provide defense for the Moscow area, which contained their seat of government. The U.S. retained the single Safeguard system installed in extreme North Dakota as protection for a portion of the ICBM launch facilities concentrated in that region. However, the U.S. chose not to deploy an additional system for the Washington, D.C., area.[18]

Testing of the PAR began in 1974 with tracking of orbiting satellites used to verify that the sensor could acquire and track up to forty objects, per its specification. These trackable objects primarily needed to have a radar cross section larger than a baseball.

The Safeguard ABM system became operational in April 1975, and 24-hour-a-day ballistic missile defense commenced at that time. This single system provided protection for that portion of the U.S. strategic deterrent capability, both manned B-52 bombers and Minuteman ICBMs, located at or in the vicinity of Grand Forks, North Dakota.

Tracking Space Objects

The sensing elements of the Space Defense Center (SDC) were called Spacetrack, a network of sensors operated by the USAF's Air Defense Command, and the U.S. Navy's Space Surveillance System (SPASUR). SPASUR consisted of three transmitters and six receivers across the southern U.S. Together the information from these sensors made it possible to keep track of all items in earth orbit. The information determined from the multiple radar returns was the position, velocity, and time for a single orbit so that future positions of each object could be calculated. These items tracked were satellites, launch vehicles, and debris left over from activities of all nations outside the earth's atmosphere. Some of these items were only in space for a short time and others for varying lengths of time, including some projected to remain in orbit for thousands of years. In the initial years of the SDC, there were typically 1000 satellites in orbit out of a total of 5000 objects being tracked. Keeping track of it all made it somewhat easier to identify that a new item had been added. The key objective was then to assess the nature of the newly added object.[19]

The Ballistic Missile Early Warning System (BMEWS), described previously, had been developed and deployed to detect and predict the impact of a ballistic missile on a path over the extreme of North America. BMEWS specifically detected a missile launched onto a ballistic path, i.e., a path not leading to an orbit. There were two additional paths for a missile attack which BMEWS did not address: the fractional orbital bombardment system (FOBS), and the multiple orbital bombardment system (MOBS). As the names imply, both of these approaches launched a weapon-carrying object into earth orbit, at least temporarily, which would subsequently release the weapons. The sensing and tracking of the Space Defense Center were able to deal with these potential approaches to missile warhead delivery.

By the late 1970s, in addition to the BMEWS and SPASUR radars, a large phased array radar was erected on the Alaskan island of Shemya called Cobra Dane, and similarly in the Florida panhandle at Eglin AFB. At Shemya, the Cobra Dane replaced the earlier FPS-17

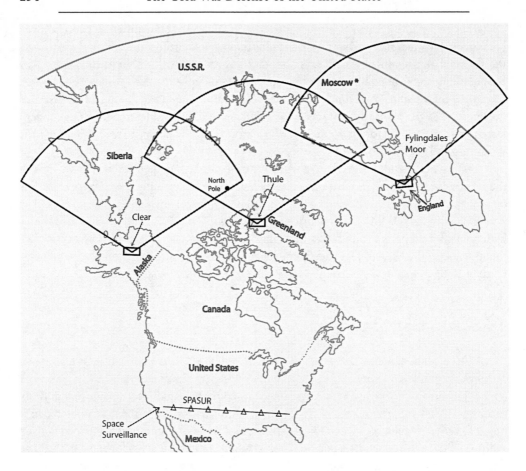

The sensing elements of NORAD's Space Defense Center.

radar in observing Soviet ICBM tests. These radars had multiple missions, one of which was to observe and report space objects. Additionally, the experimental radar of the Lincoln Laboratory at Millstone Hill, Massachusetts, could be called on when needed. A supporting asset to this tracking and assessing of orbiting space objects was a highly capable optical telescope. Located at Cold Lake, Alberta, Canada, this Baker-Nunn telescope could photograph an object the size of a basketball at 25,000 miles in space.[20]

If an object in earth orbit was judged such that its destruction was required, the U.S. could launch an anti-satellite, or ASAT, weapon. Beginning in 1963, this capability had been developed by the Air Force using the Thor IRBM rocket mated with a payload unique to the ASAT mission. Upon completion of this development, these weapons were deployed to the U.S. Territory of Johnston Island in the central Pacific. Additionally, for part of 1963 and 1964 a U.S. Army Nike Zeus missile, adapted to the anti-satellite mission, was located on the Kwajalein Atoll, also in the central Pacific. This system was kept in readiness for potential launch of a Zeus missile against an orbital object. In essence, these systems launched a nuclear payload onto a path that would bring it sufficiently close to the orbiting target so that its detonation would destroy the target. Testing leading up to the deployment of the Thor-based system confirmed that the Electro Magnetic Pulse (EMP), a significant release of electrical charge generated by a nuclear explosion, would destroy the electronics of the target missile. The launch of a weapon from either of these launch systems would have been based on the determination of the intercepting path in space derived from the data generated by the tracking and prediction of the Space Defense Center. (The developments of the Thor IRBM and the Nike Zeus anti-ballistic missile system are covered in greater detail in other sections of this book.). In 1966 an alternate payload to photograph a space object from the close pass distance provided by the intercepting Thor missile was developed. Although developed for the military application of inspecting an unknown object in space, it was also used to examine U.S. non-military satellites to determine options regarding a troubled satellite.[21]

SUBMARINE MISSILE LAUNCH DETECTION

As the threat of possible missile launch areas came to include seaward approaches to the continental U.S., beginning in 1971 several coastal radar sites were converted to detect

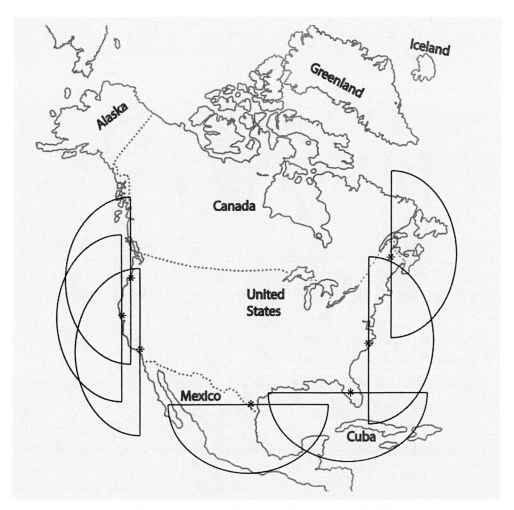

The sensing elements of the SLBM Detection and Tracking System.

missile launches from submarines. Seven radar sites on both coasts were converted to become the sensing element of the SLBM Detection and Warning System. The radars at these sites were upgraded to the FSS-7 system, which included data processing equipment to determine the track of a missile from the radar returns. The information transmitted to the Cheyenne Mountain system included location of impact ellipses, the number of incoming missiles, and the time until first impact. This information was used to create the Threat Summary Display viewable within the NORAD complex. The NORAD system relayed this information to the SAC Command Post and the National Military Command Center. These upgrades of the seven radar sites constituted a short-term measure to detect a missile launch within the line-of-sight range of the radar site. Even before completion of these short-term upgrades to the FSS-7 configuration, there was concern that the detection range achieved by this system was inadequate for the range of Soviet missiles coming into use. These systems were upgraded to achieve a range of 750 nm before becoming operational by 1972.[22]

Although detection over this range from the coastline provided some protection, the longer range of the phased array radar, as in BMEWS, was needed for Atlantic and Pacific launching areas. This concern also led to initiation of the next more capable radar system. The more capable system became known as Pave Paws and called for deployment of a

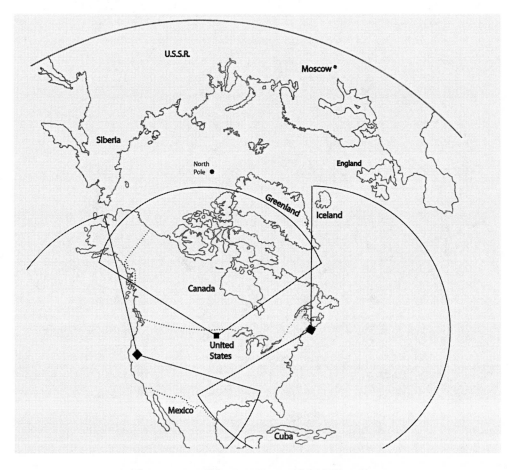

The coverage of the Pave Paws and PARCS radars.

phased array radar on each of the Atlantic and Pacific coasts of the continental U.S. With the termination of operations of the Safeguard system in 1977, the PAR radar was transferred to the Air Force, and reconfigured for the mission of monitoring Arctic Ocean areas for submarine missile launches. This site was named Perimeter Acquisition Radar Characterization System, or PARCS. The PARCS radar detected objects out to a range of 1800 nm, through an arc of 120°.

In 1980 the initial two Pave Paws radar sites, located at Otis AFB, on Cape Cod of Massachusetts, and at Beale AFB, northeast of Sacramento, California, replaced the earlier FSS-7 sites of the SLBM Detection and Warning System. These new phased array radars each had two faces, each providing 120°, for a total of 240° of coverage in azimuth. These faces were tilted 20° from the horizontal to allow vertical coverage from a few degrees to essentially 90° elevation. The radars had a range of 3000 nm, providing detection for most of the ocean areas adjacent to the North American coasts.

This set of ocean approach monitoring radars was further expanded to place similar installations at El Dorado AFS, in Texas, and at Robins AFB, in central Georgia. These stations provided backup coverage for a portion of the area covered by the initial two sites, and extended the detection area for submarine launches to the south for both coasts.[23]

With the BMEWS sites providing coverage outward from their northern perches in Alaska, Greenland, and England, and the PARCS site providing coverage of the interior of northern North America, the entire region was monitored for inbound weapons.

How NORAD Is Watching

The holy of holies within the NORAD Combat Operations Center is the Command Post, where the NORAD commander would sit and make any needed decision during an attack or other crisis. The Command Post was set up on three levels with the commander located at the middle level. Also at the mid-level were computer terminal work stations occupied by the senior battle staff officers, who were the director of the COC, the command director and his assistant. Supporting this senior staff were several technicians to operate the command and control system and to manage communications. On the lower level, just below the commander, were located additional members of the battle staff and representatives of the component military commands providing forces to NORAD. The component commands are the U.S. Air Force Air Defense Command, and the U.S. Army Air Defense Command. These officers would communicate with stations or units to obtain clarifying or amplifying information, as well as pass orders from the NORAD commander such as to launch aircraft or bring additional interceptors to a higher alert status. Behind and above the commander were located the intelligence watch officer to manage the intelligence Data Handling System, representatives of the Civil Defense National Warning Center, and the Federal Aviation Administration. All of these positions could view the pair of 16-by-20-foot screens located on the front wall of the room. One of these screens displayed a computer-generated image and the second screen the projection of a display of status information that did not require constant update.[24]

Similar in some respects to the gathering and display of information by the SAGE centers, the electronic data handling of the new NORAD operations center of 1966 was accomplished by computers receiving, storing, retrieving, displaying, and transmitting information upon the decisions of the human operators. The information received from all the sensor and defensive installations, all over the continent, is either translated to a digital format for storage in the system, or immediately displayed on a specific console dedicated for that purpose. For example, the time to first impact by an inbound missile warhead detected by BMEWS is displayed by numerical digits on a specific console, whereas the predicted impact area for that attack is stored in the computer so it can be included in a generated display.

An operator of one of the fifteen display consoles, four located in the Command Post, could build up a presentation, normally on top of an outline map of North America. Through a panel of switches, the operator would call up stored information from the computer's memory for placement relative to the map. Information from the following types could be included:

1. Tracks of suspected enemy aircraft
2. Tracks of known enemy aircraft
3. Areas of potential destruction by an attack
4. Areas of damage by chemical or biological weapons
5. Areas likely affected by radioactive fallout from an attack
6. Positions of orbiting satellites or space debris
7. Status of NORAD defensive weapons, both interceptor aircraft and missiles
8. Weather reports[25]

Having built up a display on one specific display screen, the operator could send this display to one or more of the fifteen display groups in the center. An area of this small screen display could be selected by the light pen to be enlarged by a factor of sixteen. Additionally, the display from a single console screen could be sent to the camera processor-projector unit, which captured the screen for display on the large display screen on the forward wall. To make this color display, individual symbols of the small screen version were marked by the operator as to what color they should appear on the color display. This setting establishes the intensity for that symbol in each of the red, yellow, and blue colors. The processor projector took three photographs of the screen, one for each color with the symbol at the proper intensity for that color. The three exposures were located on different sections of the film. The film was then automatically developed before being projected, each of the three images through a red, yellow, or blue filter. The resulting projected image renders each symbol in one of seven colors on the display screen. This process of transferring this individual small screen display for projection on the 16-by-20 foot screen took about ten seconds.[26]

As the NORAD system matured into the later 1970s, its mission continued to be refined. The mission was divided into three areas:

1. Provide surveillance, warning, and assessment of ballistic missile attack on North America

2. Provide monitoring and control of sovereign airspace over North America by physical inspection of unknown aircraft

3. Provide a limited defense against bomber attack on North America

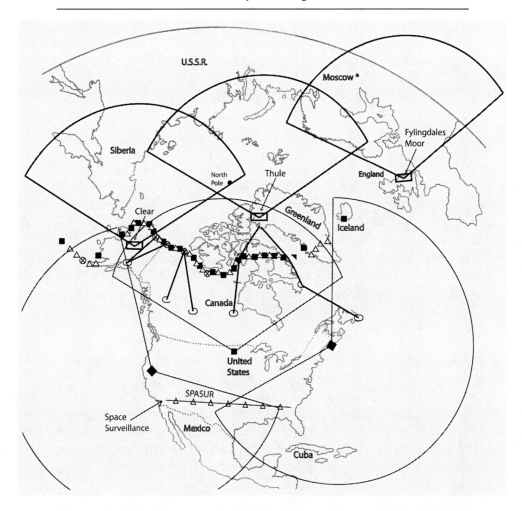

The collection of radars of NORAD searching for aircraft, missile and space activities.

If the sensing apparatus depicted above detected any suspected approaching intruder, the NORAD commander could and would dispatch one of the types of response units discussed in earlier chapters to investigate. This could be one of the types of airplane or ship discussed.

The NORAD operations center in Cheyenne Mountain has continually evolved since the early years described here. But it is most particularly the computer technology that appears most different. The principles and underlying methods of the command of the nation's defense are still in use and are relied upon absolutely today, and doubtless will be into the foreseeable future.

6

Results and Conclusions

With the long European history of a war every few years, followed by the inevitable rebuilding, then building for the next war to repeat the cycle had to be, at the least, a poor series of investments. By the end of World War II, the level of destruction achieved by the war would have meant near-complete rebuilding every few years, had the pattern continued. Although the Cold War took unimaginable funds from both sides to create the block to these periodic wars, one important thing was removed from the scene: the repetitive cycle of tearing down and subsequent rebuilding. During the period of the Cold War, after the initial rebuild, each succeeding generation was able to build on the accomplishments of their forerunners. I think this is called progress.

A Tremendous Amount of Building

At the heart of establishing the defenses presented herein was the national need or desire to defend against attack. At the start of the political tensions now called the Cold War, there was little of those defenses in existence; they had to be built, and some were invented during this time. The magnitude of the tasks involved to establish these capabilities, and the attendant costs, should be a strong testament of what the U.S. is capable of, especially when riled. This is most often called the national will.

Many of the developments presented were funded by the Congress over the initial objection of the sitting president. Presidents Truman and Eisenhower had policies of restricting defense spending following World War II; for them it was time for other spending to more directly aid the postwar population. Presidents Kennedy and Nixon initiated thorough reviews of defense programs upon taking office to ensure the vast sums spent were actually useful and necessary for the nation's protection.

Additionally, many of the developed systems were second-, third-, or fourth-generation systems of a specific type.

This evolution was required for multiple reasons. First, the initial systems deployed were often an interim capability; a more effective and permanent solution was desired. Secondly, the equipment, often continuously in use for training and preparation, simply wore out over time and needed replacement. And thirdly, technological improvements by the adversary forces required the replacement of existing equipment with more capable versions.

The following chart shows the timeframe for much of the equipment discussed in this book. It shows the number of systems under development at any point in time as well as those systems that were operational at a given point in time. What is not shown is the

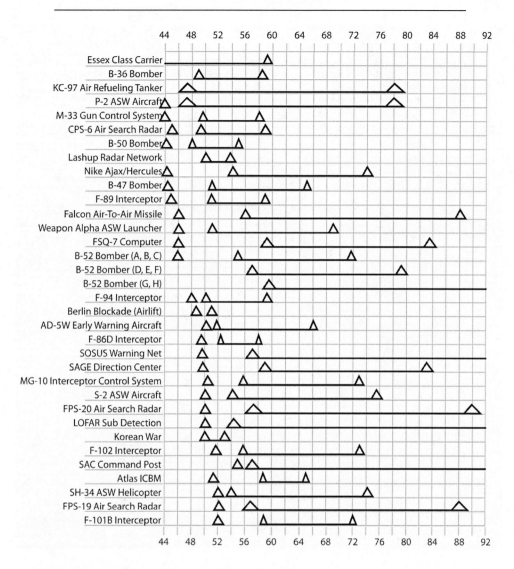

Above and following page: **Chart showing the design initiation time and the in-service time of equipment discussed in this book.**

entire origin path of each equipment, particularly for aircraft. Often a particular version for a single purpose evolved out of other work to produce a different model with a different purpose. Similarly, naval ship development had significant commonality and was largely continuous over much of this time. In this chart, the single caret is the date of the design inception, the bar terminated by two carets shows the dates of the operational history.

The chart indicates, by the number and types of different systems, the vast industrial base needed to develop and sustain such a diverse set of equipment. Each of these systems also required a logistics depot for long-term maintenance, service technical schools to provide the continuing flow of operational and maintenance personnel, and operating base facilities for the personnel near the deployment locations.

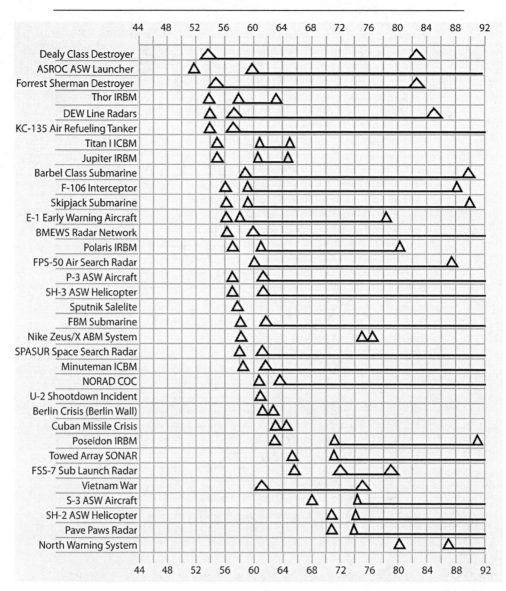

The funding required to develop these equipments, build installations, and provide the people to operate and maintain them was vast by any measure. Another testament to the will of the American people over this period of more than 40 years is that the Congress had to pass funding bills every year for those 40 years. Additionally, nine consecutive U.S. presidential administrations had to approve the funding and maintain priorities accordingly. Some of these presidents shifted those priorities somewhat, emphasizing either the discontinuance of a particular approach or the furtherance of a particular capability. But national defense remained a top priority through to the end of the Cold War.

The following chart shows the numbers of "things" purchased that are discussed in the preceding chapters. This chart is intended to show that it required a lot of equipment to be positioned around North America to provide defense. Several notes are required, however. The first thing the chart shows is that the U.S. did not purchase very many of the types

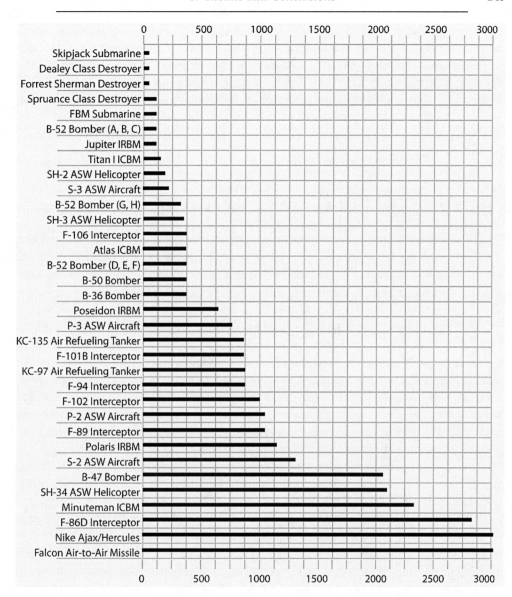

Chart showing the number of pieces purchased of equipment discussed in this book.

of naval ships discussed in the book; this is for a couple of reasons. First, specifically with destroyer ships of this period, the U.S. had significant quantities of ships built during World War II. These ships were modernized, some more than once, and continued to serve into the 1980s. Secondly, naval ship evolution is nearly continuous, which results in few truly identical copies of a design being purchased.

A second thing to note is that the Cold War period ushered in the age of jet aircraft. The early designs of the late 1940s and early 1950s were not as reliable and required more maintenance than later designs. For example, notice that the U.S. bought about 100 fewer F-94 than F-102 interceptor aircraft. Yet notice from the previous chart of operational years that the F-94 served for around 9 years while the F-102 served for 17 years. The greater reliability of the equipment is also reflected in the following chart showing the number of people serving in these areas through the period. Notice that during the 1960s and 1970s the number of people it took to provide these defenses was decreasing as the number of types of systems increased. These comments apply to the electronics as well. The ships, aircraft, sensor sites, and missiles depicted were full of electronics. The scaling of this chart was selected to depict the wide variance of numbers of different things acquired and deployed. The off-scale numbers of missiles purchased at the bottom of the chart are over 30,000 and over 40,000 respectively.

Many People Did Military Service in These Defenses

In preceding sections I have shown through descriptions and maps the vast scope of the installation needed to defend America from attack. Although the geographic spread and the number and types of installations and equipment discussed may imply the magnitude of the human element of the commitment, the following is intended to be a more direct accounting.

The vast majority of the soldiers, sailors, and airmen who served during the 40-plus years of the Cold War were young men serving a single tour as an enlisted man in one of the service branches. The most common duration was four years, which included time for basic military training as well as technical training in a specialty. The type of units described needed a high proportion of technicians trained in propulsion and electronics, in addition to operators of usually the most technical equipment in use by the military at a given time; this technical training time averaged four months, leaving approximately three and a half years to be assigned to duty stations. This created a constant flow of people to be integrated into the various operations. Such a flow led to a constant cycling of units through training. At each training cycle there was a group of people fresh from the formal training schools combined with the group with a year or so of experience now acting as the first line of supervision, combined with whatever group of transfers had arrived during the period.

The different lines in the plot show the number of people in the groups discussed in this book. They are Air Force Air Defense, Army Air Defense, Navy Air Defense, Navy ASW, SAC, and the SLBM forces. Air Force Air Defense are the troops who manned the radar sites, the network of communications sites needed to relay information to the command and control sites, the command and control sites, and the interceptor aircraft and missile units to oppose any attacking force. Army Air Defense includes the people first at the antiaircraft artillery sites and includes the people later manning the Nike missile sites. Navy Air Defense are those ships and aircraft deployed off the coastlines to extend the coverage of the ground-based radar systems. Navy ASW are the people manning shore-based detection stations, both land-based and ship-based aircraft squadrons, and crews of ships assigned to anti-submarine duties. The SAC line shows the people assigned to the strategic bomber, missile, and refueling aircraft. Similarly, the SLBM line includes the people assigned to the Navy submarine-launched ballistic missile submarines. For all lines, included are the command and control, the administration, and the maintenance personnel directly supporting these units and their equipment. What is not included are the people spread across the country in training sites, and other base personnel supporting all the activities

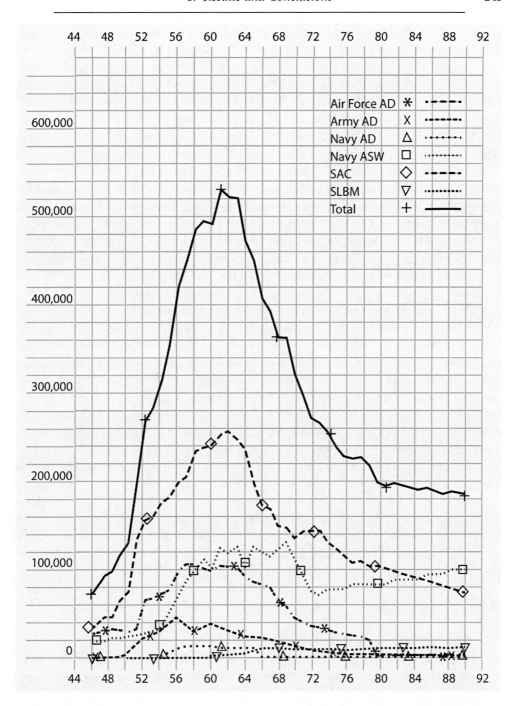

Chart showing the number of people serving in the type of defense units discussed in this book.[1]

at a particular base, such as for hospitals and police. All of the units included in the numbers would have training schools to produce the next round of manpower for these specialties, which are not shown in these numbers. In short, you could not provide these forces on a continuing basis without the existence of a lot of support from schools, bases, purchasing offices, personnel offices, and the like. Of course all of these people supported the other missions of the service branches as well. But the true number of people to provide for the defense from attack must include a portion of these supporting functions.

The line depicting Navy anti-submarine forces is a topic needing clarification. At the end of World War II and continuing, the Navy had specific units including ships and aircraft that constituted the primary ability of the U.S. to oppose enemy submarines. This continued into the Cold War with the first postwar types of aircraft carriers and their supporting fleets. However, in the 1970s, the evolution of all the equipment caused the Navy to adopt a more generic carrier battle group makeup, which was to perform all the naval operations which were based on the aircraft carrier as the principal element. Beginning with this change, the anti-submarine functions were only part of the potential duties a naval fleet might perform. Furthermore, as the naval tactics evolved, a key aspect of the anti-submarine capabilities of the carrier battle group was the protection of that carrier group from submarine attack. These factors make it more difficult to estimate the number of people assigned to protect the nation from external attack. The chart presents the total number of people with anti-submarine responsibilities, by the connection that any attack by a foreign submarine on U.S. Naval forces was really an attack on the U.S. requiring defense.

It should be noted that the SLBM submarines each had two complete crews assigned, the Blue and the Gold. While one of these crews was in the boat performing a strategic deterrent patrol, essentially the only mission of these vessels, the alternate crew was at the base for two to three months. It was during this period ashore that crew members took leave, received training, and took care of personal matters not practical to attend to while on a submarine. Prior to embarking on the next patrol of the ship, this alternate crew trained as a crew to be at the highest readiness at the time of changeover with the crew at sea. With a ready crew prepared to take over the boat when it returned from a patrol mission, the time in port was minimized and the missile boats spent far more time at sea than traditional naval ships.

It also should be noted that the numbers serving in SAC, during the times of the Korean and Vietnam Wars, were greater than the number required to maintain the deterrent forces discussed in this book. During both of these wars, SAC also provided men and equipment to conduct combat operations.

The Fall of the Soviet Communist Threat

The Cold War ended with the collapse of the Soviet Union. A common milestone marking this time is the breaching of the Berlin Wall in 1989. This barrier to human movement had been constructed by the East Germans in the late summer and fall of 1961 to prevent East Germans and others from Soviet Bloc countries from exiting the communist area to the democratic western areas. At the end, the East German border guards were simply overwhelmed by the mass of citizens who wanted to cross, and did not oppose those citizens as they tore down of part of the barrier.

Another popular image is that of U.S. President Ronald Reagan making a speech in front of the Berlin Wall in 1987, which included the oft-quoted sentence: "Mr. Gorbachev, tear down this wall." The comment was addressed to Mikhail Gorbachev, the final Premier of the

Soviet Union. Many scholars attribute President Reagan's initiative for a space-based ballistic missile defense system, commonly referred to as Star Wars after the series of popular movies of the 1970s and 1980s, as a significant factor. The massive spending of this Strategic Defense Initiative (SDI) in the 1980s was destabilizing to the degree that many observers believe that the military spending by the Soviet Union to offset or counter the SDI simply was too much, and the economic impact brought down the entire Soviet Communist system.

Technically it was a declaration passed by the Soviet of the Republics of the Supreme Soviet of the Soviet Union, declaration 142-H, that declared the fifteen former members of the Soviet Union to be independent states. This declaration occurred on December 26, 1991. The previous day, President Mikhail Gorbachev had resigned his position and declared his office extinct. The all-important control of nuclear weapons was passed to Boris Yeltsin, the president of what had become the Russian Federation on December 25, 1991.

In the preceding months, the republics of Armenia, Azerbaijan, Belarus, Estonia, Georgia, Kazakhstan, Kyrgyzstan, Latvia, Lithuania, Moldova, Russia, Tajikistan, Turkmenistan, Ukraine, and Uzbekistan had all individually seceded from the Soviet Union. Some of the former Soviet republics asked for and were granted entry to the U.N. as independent nations, whereas others retained closer ties to the Russian Federation in the Commonwealth of Independent States, a body largely intended to preserve the economic ties between the former Soviet Republics.

Over the first years of the 1990s, one by one the former Soviet satellites established governments of their own choosing, thereby attaining the lofty goal of the Atlantic Charter many years before.

Where Does That Leave Us?

These defenses worked; during the Cold War, the United States was not attacked. Furthermore, the world has gone over seventy years without another European war.

After the Cold War with the Soviet Union, it was fashionable to say that the U.S. won the Cold War. And specifically, there were assertions that SAC won the Cold War. These are in reference to the traditional concept of warfare that the army that retains the battlefield, i.e., is not driven from it, is considered the victor.

I would argue that this is incorrect in that it was not only SAC that was on the battlefield of the Cold War on the side of democracy and freedom. There were other "warring parties" present on the battlefield as well, such as those discussed in this book. In addition to the SAC presence, that of providing threat of certain destruction, there were the other defensive forces, those of preventing a successful attack on America. Additionally, there were other concepts at work, such as the U.N., the Marshall Plan, the U.S. Agency for International Development, and Radio Free Europe. The latter are only the examples of U.S. government activities to aid nations to pursue their self-determination; there would presumably be counterparts from other democratic nations. It is the combined result of all of these activities over forty-plus years' time that won the Cold War; the referenced assertions ignore the impact of these other initiatives in preventing conflict.

The individual story lines presented in this book—air defense, submarine defense, strategic deterrence, and missile defense—became stable, i.e., the rate of innovation being applied to these topics slowed down considerably, in the late 1970s and early 1980s. This is because these solutions were working; these responses were a solution to the larger problem of vulnerability to attack, and as a result the U.S. could stop looking. This combination of activities, both military and other, brought about the desired security from attack by the Soviet Union.

With the end of the Cold War, substantially with the demise of the Soviet Union, and a significantly greater degree of self-rule in the former satellite countries, the U.S. did not cease completely the preparations to defend the homeland. One of the legacies of the defense system built to defend against a Soviet attack is the continuing defense remaining today. Although much of the actual hardware purchased over the decades of the Cold War is long worn-out and obsolescent, the replacement equipment is much more capable. Of the systems discussed in this book, all have been replaced, albeit some with different capabilities.

Cold War Capability	*Modern Counterpart*
Air Defense Radars	The Pine Tree and DEW line radars have been retired. Some of the stations in Canada were upgraded to become part of the North Warning System, which is still active.
Air Defense Interceptors	Now provided by F-15, F-22, and F-18 fighters. We no longer have nuclear-armed air defense missiles.
Ground-Based Air Defense Missile	The U.S. now has mobile Patriot Missile batteries, which can be deployed when and where needed.
Long-Range Submarine Detection	The SOSUS net is significantly reduced, but we now also have ships that can deploy a roughly similar sensor system to a region such as the North Atlantic or Northwest Pacific when needed.
Airborne Anti-Submarine Response	The long-serving P-3 aircraft is still active, but is being retired in favor of the Boeing P-8 Poseidon aircraft.
Ship-Based Submarine Detection	Newer models of towed array sonar equip destroyer and cruiser ships.
Ship-Based Anti-Submarine Response	Quite similar conceptually. The missiles fired from destroyers are longer range.
Deterrence by Manned Bomber	Some B-52–H models are still active as well as newer B-1 and B-2 aircraft. All these aircraft can also carry cruise missiles.
Deterrence by Ground-Based Ballistic Missile	The U.S. has 450 Minuteman III missiles in hardened silos located in the upper Midwest. The Minuteman III carries up to three MIRVs.
Deterrence by Submarine-Launched Ballistic Missile	The Ohio Class submarines carry the Trident II D5 fleet ballistic missile, each with Multiple Independently-targetable Reentry Vehicles (MIRV).

Cold War Capability	*Modern Counterpart*
Ground-Based Ballistic Missile Defense	The U.S. now has Terminal High Altitude Area Defense (THAAD) missile batteries, which can be deployed when and where needed. These missiles intercept an incoming warhead during the downward, i.e., terminal phase of its flight.
None	Ship-Based Ballistic Missile Defense—The AEGIS combat system of modern destroyers and cruisers can also provide ballistic missile defense provided the ship is loaded with the proper missiles.

Table of modern counterparts to the type of systems discussed in this book.

After the collapse of the Soviet Union, although this left the U.S. as the sole major superpower, the U.S. received very little in the way of the long-desired peace dividend from the reduced need to maintain the level of military readiness. Perhaps the irony of the Cold War, with its high degree of focus by the U.S. on the Soviet Union—a single adversary—was that the U.S. and the other western nations had more potential adversaries to monitor, and prepare for, in the post–Cold War world. Presumably the likelihood of nuclear conflict was reduced, however.

There has been a reduction in the readiness of forces by the western nations, including the U.S. On September 27, 1991, U.S. President George H.W. Bush issued an order for much of the Strategic Air Command alert bomber force to stand down. The following day, the Secretary of Defense issued orders removing from 24-hour alert status the forty B-52 bombers that had been on constant alert for forty years, and similar orders for the much newer B-1 force. The nuclear weapons were removed to secure storage areas. President Bush issued this order in anticipation of the fall of the Soviet Union, and he challenged Soviet leader Gorbachev to respond in kind.

There is a direct parallel between this Cold War and the newer vulnerability to attack from extremists, often termed terrorists. For this newer threat, the U.S. has not yet found the combination of response that convinces would-be attackers of the futility of such an attack. A fundamental difference between the two eras is that in the late 1940s, the U.S. could identify the thing it was opposed to; it was called the Soviet Union, and it had geographic limitations and traditional military capabilities. It was necessary to prevent the military forces of the Soviet Union from successfully carrying out an attack. This degree of delineation is not present in the modern conflict. The U.S. must keep searching for and innovating toward that solution, but I am convinced that an element of that solution will be a clear demonstration of Americans' resolve to preserve their way of life.

Glossary

AA—Anti-Aircraft.

AAA—Anti-Aircraft Artillery.

AADCP—Army Air Defense Command Post.

ABM—Anti-Ballistic Missile.

ADC—Air Defense Command. A major command of the United States Air Force.

AFB—Air Force Base. A ground establishment of the United States Air Force. Larger than an Air Force Station.

AI—Airborne Intercept.

ASROC—Anti-Submarine Rocket.

ASW—Anti-Submarine Warfare.

BMEWS—Ballistic Missile Early Warning System.

DRT—Dead Reckoning Tracer.

ECM—Electronic Countermeasures.

ESM—Electronic Support Measures.

GCI—Ground Controlled Intercept.

GIUK—Greenland-Iceland-United Kingdom.

ICBM—Intercontinental Ballistic Missile.

Knot—A term of speed equaling 1 nautical mile per hour.

KT—Kiloton, equal to 1000 tons of TNT.

MAD—Magnetic Anomaly Detector. In World War II, first name as Magnetic Airborne Detector.

MT—Megaton, equal to 1,000,000 tons of TNT.

NATO—North Atlantic Treaty Organization.

NORAD—North American Air Defense Command.

NSC—National Security Council.

PPI—Plan Position Indicator.

RAF—Royal Air Force. The British armed service branch that primarily operates aircraft.

RDF—Radio Direction Finding.

SAC—Strategic Air Command. A major command of the United States Air Force.

SAM—Surface-to-Air Missile.

SAGE—Semi-Automatic Ground Environment.

SLBM—Submarine-Launched Ballistic Missile.

SOSUS—Sound Surveillance System.

UN—United Nations.

USAAC—United States Army Air Corps.

USAFE—United States Air Force Europe. A major command of the United States Air Force.

USSR—Union of Soviet Socialist Republics. Also known as the Soviet Union.

Historic Sites

Historic sites in the United States that represent events of the Cold War are less numerous than such sites for World War II, for instance. In the research for this book the author visited the following sites. A small number of these sites require arrangements in advance because they are located on an active military installation. However, you can still visit.

Aerospace Museum of California, 3200 Freedom Park Drive, McClellan, California 95652

Bradbury Science Museum, 1350 Central Avenue, Los Alamos, New Mexico 87544

Castle AFB Museum, 5050 Santa Fe Dr., Atwater, California 95301

Comox Air Force Museum, Military Row, Lazo, British Columbia VOR 2KO, Canada

Computer History Museum, 1401 N. Shoreline Blvd., Mountain View, California 94043

Evergreen Aviation & Space Museum, 500 NE Captain Michael King Smith Way, McMinnville, Oregon 97128

Hill Aerospace Museum, 7961 Wardleigh Road, Hill AFB, Utah 84056

Malmstrom Museum and Air Park, 90 Whitehall Drive, Malmstrom AFB, Montana 59402

March Field Museum, 22550 Van Buren Blvd., Riverside, California 92518

Minuteman Missile National Historic Site, 24545 CottonWood Road, Philip, South Dakota 57567

Moffett Field Historical Society Museum, 126 Severyns Ave., Mountain View, California 94043

Museum of Flight, 9404 E. Marginal Way S., Seattle, Washington 98108

National Air and Space Museum, 600 Independence Ave. SW, Washington, D.C. 20560

National Museum of Nuclear Science & History, 601 Eubank Blvd SE, Albuquerque, New Mexico 87123

National Museum of the United States Air Force, 1100 Spaatz St., Dayton, Ohio 45431

National Naval Aviation Museum, 1750 Radford Blvd., NAS Pensacola, Florida 32508

Nike Missile Site SF-88, Mill Valley, California 94941

Oakland Aviation Museum, 8252 Earhart Rd., Oakland, California 94621

Palm Springs Air Museum, 745 N. Gene Autry Trail, Palm Springs, California 92262

Peterson Air and Space Museum, 150 Ent Ave., Peterson AFB, Colorado 80914

Pima Air and Space Museum, 6000 E. Valencia Road, Tucson, Arizona 85706

Planes of Fame Air Museum, 14998 Cal Aero Drive, Chino, California 91710

Planes of Fame Air Museum, 755 S. Mustang Blvd, Williams, Arizona 86046

Ronald Reagan Minuteman Missile State Historic Site, 555 113½ Ave. NE Hwy. 45, Cooperstown, North Dakota 58425

South Dakota Air and Space Museum, 2890 Davis Drive, Ellsworth AFB, South Dakota 57706

Strategic Air Command & Space Museum, 28210 W. Park Hwy., Ashland, Nebraska 68003

Titan Missile Museum, 1580 W. Duval Mine Road, Green Valley, Arizona 85614

Travis Air Force Base Heritage Center, Building 80, 461 Burgan Blvd., Travis AFB, California 94535

USS *Blueback*, 1945 SE Water Ave., Portland, Oregon 97214

USS *Hornet* Museum, 707 W. Hornet Ave., Alameda, California 94501

USS *Iowa* BB-61, 250 South Harbor Blvd., San Pedro, California 90731 (Long Beach Harbor)

USS *Midway* Museum, 910 N. Harbor Dr., San Diego, California 92101

USS *Turner Joy*, 300 Washington Beach Ave., Bremerton, Washington 98337

USS *Yorktown* CV-10, 40 Patriots Point Rd., Mt Pleasant, South Carolina 29464 (Charleston Harbor)

Vandenberg AFB Space Launch Complex 10, Vandenberg AFB, Lompoc, California 93437

White Sands Missile Range Museum, Wsmr P. Rt. 1, White Sands Missile Range, New Mexico 88002

Wings Over the Rockies Air and Space Museum, 7711 E. Academy Blvd. #1, Denver, Colorado 80230

Yanks Air Museum, 7000 Merrill Ave #35-A270, Chino, California 91710

Chapter Notes

Chapter 1

1. The Yalta Conference, avalon.law.yale.edu/wwii/yalta.asp, retrieved 5/20/2018.
2. "History of Strategic Air and Ballistic Missile Defense," vol. 1: 1945–1955" (Center for Military History, United States Army, Washington, D.C., 1975), 44.
3. Andrei Cherny, *The Candy Bombers* (New York: Penguin, 2008), 80.
4. Fulton, Missouri 1946. winstonchurchill.org/resource/speeches/1946-1963-elder-statesman/the-sinews-of-peace/, retrieved 5/20/2018.
5. Truman Doctrine. avalon.law.yale.edu/20th_century/trudoc.asp, retrieved 2/20/2017.
6. 1948 Czechoslovakia coup d'état. en.widipedia.org/wiki/1948_Czechoslovakia_coup_d%27%C3%A9tat, retrieved 8/1/2017.
7. Cherny, 241.
8. Cherny, 256–259.
9. "U.S. Objectives with Respect to the USSR to Counter Soviet Threats to U.S. Security," NSC 20/4, November 23, 1948, 3.
10. "United States Objectives and Programs for National Security," NSC 68, April 14, 1950, 1–47.

Chapter 2

1. Jack Gough, *Watching the Skies: The History of Ground Radar in the Air Defense of the United Kingdom* (Ministry of Defense Air Historical Branch, HMSO, 1993), 2–28.
2. Wikipedia contributors, "Kammhuber Line," https://en.wikipedia.org/w/index.php?title=Kammhuber_Line&oldid=835123905 (accessed May 26, 2014).
3. Gough, 20.
4. David F. Winkler, *Searching the Skies: The Legacy of the United States Cold War Defense Radar Program* (Langley AFB, VA: United States Air Force Air Combat Command, 1997), 16–18.
5. Winkler, 18–25.
6. Derek Wood with Derek Dempster, *The Narrow Margin: The Battle of Britain and the Rise of Air Power 1930–40* (New York: Paperback Library, 1969), 76–77.
7. Kenneth Schaffel, *The Emerging Shield: The Air Force and the Evolution of Continental Defense 1945–1960* (Washington, D.C.: Office of Air Force History United States Air Force, 1991), 76–80.
8. Lloyd H. Cornett Jr. and Mildred W. Johnson, *A Handbook of Aerospace Defense Organization, 1946–1980* (Peterson AFB, CO: Office of History, Aerospace Defense Center, 1980), 113–131.
9. "Radar Bulletin No. 8-A (RADEIGHTA) Aircraft Control Manual," Chapters 12 and 13, Chief of Naval Operations, 1951, 104–119.
10. *Ibid.*, 120–125.
11. "Pilot's Operating Manual for Airborne Radar AN/APS-6 Series for Night Fighters CO NAVAER 08–05–120" (Washington, D.C.: United States Navy Department, 1944), 15.
12. "Radar Bulletin No. 8-A," 127–140.
13. Mark A. Berhow and Mark L. Morgan, *Rings of Supersonic Steel: Air Defenses of the United States Army 1950–1979* (Bodega Bay, CA: Fort MacArthur Military Press, 2002), 5.
14. Schaffel, 147.
15. *The Steel Ring*. U.S. Army, *The Big Picture*, TV.
16. Berhow and Morgan, 8.
17. Berhow and Morgan, 45–178.
18. "FM 44-10 U.S. Army Air Defense Fire Distribution System AN/FSG-1 (Missile Master)," Headquarters, Department of the Army, Washington, D.C., 1963, 4–16.
19. *Army Air Defense Command (ARADCOM): The Inner Ring*, U.S. Army, *The Big Picture*, TV.
20. Schaffel, 159–160.
21. Schaffel, 210–216.
22. Captain Joseph F. Bouchard, "Guarding the Cold War Ramparts: The U.S. Navy's Role in Continental Air Defense," *Naval War College Review*, Summer 1999, 4–12.
23. Kent C. Redmond and Thomas M. Smith, *From Whirlwind to MITRE: The R&D Story of the SAGE Air Defense Computer* (Cambridge, MA: MIT Press, 2000), 283–301.
24. R.R. Everett, C.A. Zraket, and H.D. Bennington, "SAGE: A Data Processing System for Air Defense," Proceedings of the Eastern Computer Conference, May 6–8, 1958, Los Angeles, 150–153.
25. W.A. Ogletree, H.W. Taylor, E.W. Veitch, and J. Wylen, "AN/FST-2 Radar-Processing Equipment for SAGE," Proceedings of the Eastern Computer Conference, May 6–8, 1958, Los Angeles, 157–160.
26. *In Your Defense*. Dir. unknown. United States Air Force, 1963. Film.
27. *The MG-10 System: The Aircraft and Weapon Control System of the F-102A (F-0582)*. Dir. unknown. Hughes Aircraft. Film.
28. Handbook of Aerospace Defense Organization, 113–131.

29. Schaffel, 236–238.

30. "The SAGE/BOMARC Air Defense Weapons System—Fact Sheet," International Business Machines Corporation, Military Products Division, New York, NY, 1958, 47–54.

31. "Flight Training Instruction—Air to Air Intercept Procedures Workbook," NAS Corpus Christi, Texas: United States Navy Naval Air Training Command, 2010, 13–1 through 13–5.

Chapter 3

1. J.R. Hill, *Anti-Submarine Warfare*, 2nd ed. (Annapolis, MD: Naval Institute Press, 1989), 21, 27; David Miller, *An Illustrated Guide to Modern Sub Hunters* (New York: Arco, 1984), 8.

2. Bruce Hampton Franklin, *The Buckley Class Destroyer Escorts* (Annapolis, MD: Naval Institute Press, 1999), 3.

3. Alfred Price, *Aircraft versus Submarine* (Annapolis, MD: Naval Institute Press, 1973), 204.

4. "United States Fleet Anti-Submarine and Escort of Convoy Instructions, Part I: Surface Craft," United States Fleet Headquarters of the Commander In Chief, Navy Department, Washington, D.C., May 1945, 1–3 through 1–13.

5. *Ibid.*, 1–55, 1–56.

6. *Ibid.*, 1–4.

7. *Ibid.*, 1–8.

8. *Ibid.*, 1–13.

9. *Ibid.*, 1–55.

10. *Ibid.*, 1–60.

11. *Ibid.*, 1–61.

12. *Ibid.*, 1–65, 1–66.

13. *Ibid.*, 1–69, 1–70.

14. "United States Fleet Anti-Submarine and Escort of Convoy Instructions, Part II: Aircraft," United States Fleet Headquarters of the Commander In Chief, Navy Department, Washington, D.C., May 1945, 2–1 through 2–3.

15. *Ibid.*, 2–5.

16. *Ibid.*, 2–7.

17. *Ibid.*, 2–10.

18. *Ibid.*, 2–13.

19. *Ibid.*, 2–4.

20. *Ibid.*, 2–24.

21. *Ibid.*, 2–25.

22. *Ibid.*, 2–25.

23. *Ibid.*, 2–26.

24. *Ibid.*, 2–26.

25. *Ibid.*, 2–26.

26. *Ibid.*, 2–27.

27. *Ibid.*, 2–20.

28. *Ibid.*, 2–17.

29. *Ibid.*, 2–18.

30. Hill, 19.

31. NAVFAC Station History. https://www.navyhistory.com/NAVFAC Station History.txt, retrieved 6/5/2016, 2.

32. Oppenheim, Alan V., ed. *Applications of Digital Signal Processing* (Englewood Cliffs, NJ: Prentice Hall, 1978), 406–416.

33. Hill, 44–51.

34. *The Hunter Killers.* U.S. Navy film, 1967.

35. Price, 243–244.

36. Edward M. Brittingham, Capt. U.S. Navy (retired), *Sub Chaser: The Story of a Navy VP NFO* (Richmond, VA: Ashcraft Enterprises, 1999), 31–32.

37. Roger A. Holler, Arthur W. Horbach, and James F. McEachern, *The Ears of ASW: A History of U.S. Navy Sonobuoys* (Warminster, PA: Navmar Applied Sciences Corporation, 2008), 50.

38. *The Hunter Killers.*

39. *Goblin on the Doorstep.* Grumman Aircraft Engineering Company, film, 1963.

40. Brad Elward, "Lockheed Martin P-3 Orion—U.S. Service," *World Air Power Journal* 43 (Winter 2000): 100–105.

41. Peter A. Huchthausen, *October Fury* (Hoboken, NJ: John Wiley & Sons, 2002), 131–132.

42. Hill, 19–35; Norman Friedman, *U.S. Destroyers: An Illustrated Design History* (Annapolis, MD: Naval Institute Press, 1982), 193–201, 255–290; Jim Sullivan, *Aircraft Number 68: P2V Neptune In Action* (Carrollton, Texas: Squadron/Signal, 1985), 27; Elward, 59–88.

43. Huchthausen, 63.

44. *Coordinated ASW.* Dir. unknown. U.S. Navy, 1976. Film.

45. "NATOPS Flight Manual Navy Model SP-2H Aircraft (NAVAIR 01–75EEB-1)," United States Navy Naval Air Systems Command, Washington, D.C., 1969, 1–110 through 1–113.

46. *Ibid.*, 1–117 through 1–122.

47. Richard S. Dann and Rick Burgess, *Aircraft Number 193–P-3 Orion In Action* (Carrollton, Texas: Squadron/Signal, 2004), 8, 13, 16.

48. "Lockheed Martin P-3," 66–68.

49. Holler, Horbach, and McEachern, 77–79.

50. Elward, 80–81.

51. *Ibid.*, 78.

52. Friedman, 198–199, 249–251.

53. Stanley G. Lemon, "Towed Array History, 1917–2003," *IEEE Journal of Oceanic Engineering* 29, Issue 2 (April 2004): 365–371.

54. Friedman, 280–290.

55. *Coordinated ASW.*

Chapter 4

1. "The Air & Space Power Course—Airpower Theory," College of Aerospace Doctrine, Research and Education, Maxwell AFB, AL, 3–5.

2. Peter R. Faber, Lt. Colonel, "The Development of U.S. Strategic Bombing Doctrine in the Interwar Years: Moral and Legal?" *Journal of Legal Studies* (1996/1997): 4.

3. Michael S. Sherry, *The Rise of American Airpower: The Creation of Armageddon* (New Haven, CT: Yale University Press, 1987), 91.

4. General Curtis E. LeMay with MacKinlay Kantor, *Mission with LeMay* (Garden City, NY: Doubleday, 1965), 148–152.

5. LeMay, 213.

6. LeMay, 231.

7. *Target for Today (T.F. I-3384).* War Department film, 1944.

8. Jeffrey Ethell and Alfred Price, *Target Berlin—*

Mission 250: 6 March 1944 (New York: Jane's, 1981), 28–29.

9. "Bombardier's Information File," Headquarters Army Air Forces, Washington, DC, March 1945, 2-1-1 through 2-1-4.

10. *Ibid.*, 2-2-1 through 2-2-2.

11. *Norden Bombsight: Operation.* Army Air Force Training Command film.

12. LeMay, 431.

13. LeMay, 444.

14. Don Logan, *The Boeing C-135 Series: Stratotanker, Stratolifter, and Other Variants* (Atglen, PA: Schiffer, 1998), 10.

15. Mike Hill, John M. Campbell, and Donna Campbell, *Peace Was Their Profession—SAC: A Tribute* (Atglen, PA: Schiffer, 1995). 18–19.

16. List of USAF Bomb Wings Assigned to Strategic Air Command. https://en.wikipedia.org/wiki/List_of_USAF_Bomb_Wings_assigned_to_Strategic_Air_Command., retrieved 6/16/2015.

17. Chris Adams, *Inside the Cold War: A Cold Warrior's Reflections* (Maxwell Air Force Base, AL: Air University Press, 1999), 6.

18. Adams, 1–2.

19. Howard Blum, "At a SAC Base, Living Centers on State of Alert," *New York Times*, February 21, 1984.

20. *The Doomsday Bombers (SAC).* Film.

21. *SAC Command Post.* Dir. unknown. United States Air Force, 1964. Film.

22. List of USAF Bomb Wings Assigned to Strategic Air Command.

23. "P-3 Orion Inertial Navigation System," *Lockheed Orion Service Digest*, Issue 12 (December 1966): 10–19.

24. AGM-28 Hound Dog. https://en.wikipedia.org/wiki/AGM-28_Hound_Dog., retrieved 10/9/2016.

25. *Operation Headstart: United States Air Force Film Report.* Dir. unknown. United States Air Force, 1959. Film.

26. "Development of Airborne Armament 1910–1961, Vol. 1: Bombing Systems," Aeronautical Systems Division, Air Force Systems Command, U.S. Air Force, October 1961, 16–33.

27. Karl J. Eschemann, *Linebacker: The Untold Story of Air Raids on North Vietnam* (New York: Ballantine Books, 1989), 162–166.

28. Jacob Neufeld, *The Development of Ballistic Missiles in the United States Air Force 1945–1960* (Washington, D.C.: Office of Air Force History United States Air Force), 1990.

29. *Ibid.*, 35–36.

30. *Ibid.*, 73–75.

31. *Ibid.*, 122–125.

32. *Ibid.*, 152.

33. *Ibid.*, 186–193.

34. *Ibid.*, 222–225.

35. *Ibid.*, 163, 182.

36. "Lockheed Missiles & Space Company: The Early Years," *Lockheed Horizons*, Issue 12 (June 1983): 8–9.

37. *Ibid.*, 10–13.

38. Neufeld, 237.

39. J.W. Cornelisse, H.F.R. Schoyer, and K.F. Wakker, *Rocket Propulsion and Spaceflight Dynamics*, Part 1 (London, UK: Pitman, 1979), 282–288.

Chapter 5

1. David F. Winkler, *Searching the Skies: The Legacy of the United States Cold War Defense Radar Program* (Langley AFB, VA: United States Air Force Air Combat Command, 1997), 28.

2. Kenneth Schaffel, *The Emerging Shield: The Air Force and the Evolution of Continental Defense 1945–1960* (Washington, D.C.: Office of Air Force History United States Air Force, 1991), 241–246.

3. Schaffel, 246–254.

4. Marion Talmadge and Iris Gilmore, *NORAD: The North American Air Defense Command* (New York: Dodd, Mead & Co., 1967), 57–58.

5. Talmadge and Gilmore, 47–56.

6. Stanley G. Zabetakis, and John F. Peterson, "The Diyarbakir Radar," unknown CIA Publication 8 (Fall 1964): 42.

7. *Ibid.*, 46–47.

8. Melvin L. Stone and Gerald P. Banner, "Radars for the Detection and Tracking of Ballistic Missiles, Satellites, and Planets," *Lincoln Laboratory Journal* 12, Issue 2 (2000): 218–220.

9. *Ibid.*, 222–225.

10. Schaffel, 256–260.

11. Stone and Banner, 220–221.

12. *Ibid.*, 221.

13. "Ballistic Missile Defense Technologies," Office of Technology Assessment, Princeton University Press, Princeton, NJ, 1986, 45–49.

14. *A 20-Year History of the Antiballistic Missile Program.* Dir. unknown. U.S. Army Ballistic Missile Defense Systems Command, 1976. Film; *Nike-X: ABM "Army Air Defense Command" (ARADCOM).* U.S. Army, *The Big Picture*, TV, 1966.

15. Ashton B. Carter and David N. Schwartz, eds., *Ballistic Missile Defense* (Washington, D.C.: Brookings Institution Press, 1984), 333–336.

16. *Ibid.*, 336–340.

17. *Ibid.*, 341.

18. "Ballistic Missile Defense Technologies," 49–55.

19. Talmadge and Gilmore, 39–44.

20. *Ibid.*, 41–42.

21. Clayton K.S. Chun, *Shooting Down a Star: Program 437, The U.S. Nuclear ASAT System and Present-Day Copycat Killers* (Maxwell Air Force Base, AL: Air University Press, 2000), 4–18.

22. Winkler, 53–54.

23. *Ibid.*, 54–56.

24. Talmadge and Gilmore, 16–22.

25. *NORAD Command Briefing*, United States Air Force film.

26. Talmadge and Gilmore, 23–35.

Chapter 6

1. "Army Air Forces Statistical Digest 1946," Headquarters Army Air Forces, Washington, DC, June 1947, 27.

"Department of the Army Historical Summary Fiscal Year 1969," Center of Military History, United States Army, Washington, DC, 1973, 13. Identically titled documents for the years 1970–1976. For: 1970, 49; 1971, 9; 1972, 10; 1973, 14–15; 1974, 8–9; 1975, 9; and 1976, 21.

"United States Air Force Statistical Digest 1947," Headquarters U.S. Air Force, Washington, DC, August 1948, 46. Identically titled document for the year 1948. For: 1948, 14.

"United States Air Force Statistical Digest Jan 1949–Jun 1950," Headquarters U.S. Air Force, Washington, DC, April 1951, 29.

"United States Air Force Statistical Digest Fiscal Year 1951," Headquarters U.S. Air Force, Washington, DC, November 1952. p. 425. Identically titled documents for the years 1952–1980. For: 1952, 372; 1953, 362; 1954, 230; 1955, 270; 1956, 261; 1957, 267; 1958, 253; 1959, 247; 1960, 212; 1961, 234; 1962, 227; 1963, 245; 1964, 210; 1965, 194; 1966, 250; 1967, 253; 1968, 271; 1969, 237; 1970, 220; 1971, 234; 1972, 271; 1973, 117; 1974, 92; 1975, 118; 1976, 168; 1977, 125; 1978, 111; 1979, 121; and 1980, 125.

Bibliography

Adams, Chris. *Inside the Cold War: A Cold Warrior's Reflections.* Maxwell AFB, AL: Air University Press, 1999.

Aerial Anti-Submarine Warfare—Weapons Selection and Methods of Attack. United States Navy film, 1955..

Aerospace Communications—The Reins of Command. United States Air Force film, 1961.

"The Air & Space Power Course—Airpower Theory." College of Aerospace Doctrine, Research and Education, Maxwell AFB, AL, year unknown.

"Air Defense Command and Control Operations." Air Force Instruction 13–1AD, Volume 3, U.S. Air Force, June 2009.

"Air Defense Doctrine and Procedures." Chapter 2, USA ADS Digest, 1965.

Aleutian Skywatch. United States Air Force film.

"Anti-Satellite Weapons, Countermeasures, and Arms Control." Office of Technology Assessment, Princeton University Press, Princeton, NJ, 1986.

"Anti-Submarine Warfare, Including: AEGIS Combat System, SOSUS, Thermocline, Sonobuoy, SOFAR Channel, SONAR 2087, Naval Tactical Data System, SSDS, SIFOREX, Bathythermograph, Expendable Bathythermograph, and Command-Activated Sonobuoy System." Hephaestus Books, 2012. [A hardcopy of a number of Wikipedia articles.]

Army Air Defense Command (ARADCOM): The Inner Ring. U.S. Army, The Big Picture TV.

"Army Air Forces Statistical Digest 1946." Headquarters Army Air Forces, Washington, D.C., June 1947.

AWACS (Segment of Air Force Now #Unknown). United States Air Force film, December 1983.

Baker, David. *Anti-Submarine Warfare.* Vero Beach, FL: Rourke Enterprises, 1989.

"Ballistic Missile Defense Technologies." Office of Technology Assessment, Princeton University Press, Princeton, NJ, 1986.

Bath, David W., ed. *Air Force Missileers and the Cuban Missile Crisis.* Breckenridge, Colorado: Association of Air Force Missileers, 2012.

Berhow, Mark A., and Mark L. Morgan. *Rings of Supersonic Steel: Air Defenses of the United States Army 1950–1979.* Bodega Bay, CA: Fort MacArthur Military Press, 2002.

Birdsall, Steve. *Flying Colors: B-17 Flying Fortress in Color.* Carrollton, TX: Squadron/Signal, 1986.

Blackman, Tony. *NIMROD: Rise and Fall.* London: Grub Street, 2013.

Blum, Howard. "At a SAC Base: Living Centers on State of Alert." *New York Times,* February 21, 1984.

"Bombardier's Information File." Headquarters Army Air Forces, Washington, D.C., March 1945.

Bouchard, Captain Joseph F. "Guarding the Cold War Ramparts: The U.S. Navy's Role in Continental Air Defense." *Naval War College Review,* Summer 1999.

Bowman, Martin W. *Stratofortress: The Story of the B-52.* Barnsley, South Yorkshire, Great Britain: Pen & Sword, 2005.

Boyne, Walter J. "The Man Who Built the Missiles." *Air Force Magazine,* October 2000.

Brittingham, Edward M. Capt. U.S. Navy (retired). *Operation Poppy.* New York: TFG Press, 2003.

_____. *Sub Chaser: The Story of a Navy VP NFO.* Richmond, VA: Ashcraft Enterprises, 1999.

Brugioni, Dino A. *Eyeball to Eyeball: The Inside Story of The Cuban Missile Crisis.* New York: Random House, 1991.

Burns, Thomas S. *The Secret War for the Ocean Depths: Soviet-American Rivalry for Mastery of the Seas.* New York: Rawson Associates, 1978.

"Can We Defend Our Coasts Against Russian Subs?" *Popular Science,* September 1957.

Carter, Ashton B., and David N. Schwartz, eds. *Ballistic Missile Defense,* Washington, D.C.: Brookings Institution Press, 1984.

"The Changing Navigation Picture." *Flight,* April 12, 1957, 474–483.

Charo, Arthur. *Continental Air Defense: A Neglected Dimension of Strategic Defense.* Lanham, MD: University Press of America, 1990.

Cherny, Andrei. *The Candy Bombers.* New York: Penguin Group, 2008.

Chun, Clayton K.S. *Shooting Down a Star: Program 437, the US Nuclear ASAT System and Present Day Copycat Killers.* Maxwell AFB, AL: Air University Press, 2000.

Clancy, Tom. *Submarine: A Guided Tour Inside a Nuclear Warship.* New York: Berkley Books, 1993.

Colorado Experience: NORAD. Rocky Mountain PBS TV, 2014.

Coordinated ASW. U.S. Navy film, 1976.

Cornelisse, J.W., H.F.R. Schoyer, and K.F. Wakker. *Rocket Propulsion and Spaceflight Dynamics Part 1.* London, UK: Pitman Publishing, 1979.

Cornett, Lloyd H. Jr., and Mildred W. Johnson. *A Handbook of Aerospace Defense Organization—1946-1980.* Peterson AFB, CO: Office of History, Aerospace Defense Center, 1980.

Crabtree, James D. *On Air Defense.* Westport, CT: Praeger, 1994.

Cross, Robin. *The Bombers: The Illustrated Story of Offensive Strategy and Tactics in the Twentieth Century.* New York: Macmillan, 1987.

The Cutting Edge. Lockheed California Company film.

Daniel, Donald C. *Antisubmarine Warfare and Superpower Strategic Stability.* London: Macmillan, 1986.

Dann, Richard S., and Rick Burgess. *Aircraft Number 193—P-3 Orion In Action.* Carrollton, TX: Squadron/Signal, 2004.

Davis, Larry. "67th TRW in Korea." *International Air Power Review* 10, Autumn/Fall 2003.

"Defense Acquisitions: Evaluation of Navy's Anti-Submarine Warfare Assessment." United Sates Government Accountability Office, National Security and International Affairs Division, 1999.

"Department of the Army Historical Summary Fiscal Year 1969." Center of Military History, United States Army, Washington, D.C., 1973.

"Department of the Army Historical Summary Fiscal Year 1970." Center of Military History, United States Army, Washington, D.C., January 1973.

"Department of the Army Historical Summary Fiscal Year 1971." Center of Military History, United States Army, Washington, D.C., February 1973.

"Department of the Army Historical Summary Fiscal Year 1972." Center of Military History, United States Army, Washington, D.C., April 1974.

"Department of the Army Historical Summary Fiscal Year 1973." Center of Military History, United States Army, Washington, D.C., 1977.

"Department of the Army Historical Summary Fiscal Year 1974." Center of Military History, United States Army, Washington, D.C., March 1977.

"Department of the Army Historical Summary Fiscal Year 1975." Center of Military History, United States Army, Washington, D.C., 1978.

"Department of the Army Historical Summary Fiscal Year 1976." Center of Military History, United States Army, Washington, D.C., 1977.

Development of the Soviet Ballistic Missile Threat—United States Air Force Film Report. United States Air Force film, 1960.

Donald, David, ed. *Warplanes of the Fleet.* Norwalk, Connecticut: AIRtime, 2004.

The Doomsday Bombers (SAC) film.

Ecker, Capt. William B. USN (Ret), and Kenneth V. Jack. *Blue Moon Over Cuba.* Long Island City, NY: Osprey, 2012.

Elite Professionals—Today's USAF Missileers. The Travel Channel.

Elward, Brad. "Lockheed Martin P-3 Orion: US Service." *World Air Power Journal* 43, Winter 2000.

Eschemann, Karl J. *Linebacker: The Untold Story of Air Raids on North Vietnam.* New York: Ballantine Books, 1989.

Ethell, Jeffrey, and Alfred Price. *Target Berlin: Mission 250, 6 March 1944.* New York: Jane's, 1981.

Everett, R.R., C.A. Zraket, H.D. Bennington. "SAGE—A Data Processing System for Air Defense." Proceedings of the Eastern Computer Conference, May 6–8, 1958, Los Angeles, 145–155.

Exploring an Atlas Missile Silo. Hidden Hometown, video.

Eyes on the Cape (Segment of Air Force Now #146). United States Air Force film.

Faber, Lt. Colonel Peter R. "The Development of US Strategic Bombing Doctrine in the Interwar Years: Moral and Legal?" *Journal of Legal Studies,* 1996/1997.

"The Fleet Ballistic Missile System." Burbank, CA: Lockheed Horizons, Lockheed Corporation, Spring 1980.

"Flight Training Instruction—Air to Air Intercept Procedures Workbook." United States Navy Naval Air Training Command, NAS Corpus Christi, TX, 2010.

"FM 44–10 U.S. Army Air Defense Fire Distribution System AN/FSG-1 (Missile Master)." Headquarters, Department of the Army, Washington, D.C., 1963.

Franklin, Bruce Hampton. *The Buckley Class Destroyer Escorts.* Annapolis, MD: Naval Institute Press, 1999.

Frieden, David R., Lt. Commander U.S. Navy, ed. *Principles of Naval Weapons Systems.* Annapolis, MD: Naval Institute Press, 1985.

Friedman, Norman. *The Naval Institute Guide to World Naval Weapons Systems.* Annapolis, MD: Naval Institute Press, 1989.

_____. *U.S. Destroyers: An Illustrated Design History.* Annapolis, MD: Naval Institute Press, 1982.

Friedman, Richard S., Lt. Col. David Miller, Doug Richardson, Bill Gunston, David Hobbs, and Max Walmer. *Advanced Technology Warfare.* New York: Crown/Harmony, 1985.

Ft. Bliss—The Heart of Army Air Defense. U.S. Army, *The Big Picture,* TV, 1972.

Gates, Robert M. *From the Shadows.* New York: Simon & Schuster Paperbacks, 1996.

Gerken, Louis. *ASW versus Submarine Technology Battle.* Chula Vista, CA: American Scientific Corp., 1986.

"The Goblin Killers." *Time* 72, Issue 9 (September 1957).

Goblin on the Doorstep. Grumman Aircraft Engineering Company film, 1963.

Goldberg, Alfred, ed. *A History of the United States Air Force.* New York: Arno Press, 1974.

Gough, Jack. *Watching the Skies: The History of Ground Radar in the Air Defense of the United Kingdom.* Ministry of Defense Air Historical Branch, HMSO, 1993.

Gunston, Bill. *An Illustrated Guide to Modern Bombers.* London, UK: Salamander, 1988.

_____. *Night Fighters: A Development and Combat History.* New York: Scribner's, 1976.

Gunston, Bill, and Lindsay Peacock. *Fighter Missions: Modern Air Combat, The View from the Cockpit.* New York: Crown, 1989.

Hallowell, CDR Paul E. *Maritime Patrol Aircraft: Operational Versatility from the Sea.* National War College, Fort Lesley J. McNair, Washington, D.C., 1994.

Halvorsen, Gail S. *The Berlin Candy Bomber.* Springville, UT: Horizon, 2010.

Higham, Robin, and Abigail T. Siddall, eds. *Flying Combat Aircraft of the USAAF-USAF.* Ames: Iowa State University Press, 1975.

Higham, Robin, and Carol Williams, eds. *Flying Combat Aircraft of the USAAF-USAF,* vol. 2. Ames: Iowa State University Press, 1978.

Hill, J.R. *Anti-Submarine Warfare,* 2nd ed. Annapolis, MD: Naval Institute Press, 1989.

Hill, Mike, John M. Campbell, and Donna Campbell. *Peace Was Their Profession—SAC: A Tribute.* Atglen, PA: Schiffer, 1995.

History of Strategic Air and Ballistic Missile Defense, vol.

1: 1945–1955. Center for Military History, United States Army, Washington, D.C., 1975.

Hogg, Ian V. *Anti-Aircraft: A History of Air Defence.* London: Macdonald and Jane's, 1978.

Holler, Roger A., Arthur W. Horbach, and James F. McEachern. *The Ears of ASW: A History of U.S. Navy Sonobuoys.* Warminster, PA: Navmar Applied Sciences Corporation, 2008.

Huchthausen, Peter A. *October Fury.* Hoboken, NJ: John Wiley & Sons, 2002.

The Hunter Killers. U.S. Navy film, 1967.

In Your Defense. United States Air Force film, 1963.

Jockel, Joseph T. *No Boundary Upstairs: Canada, the United States, and the Origins of North American Air Defence 1948–1958.* Vancouver: University of British Columbia Press, 1987.

Katz, Kenneth P. Jim. *B-52G/H Stratofortress In Action.* Carrollton, TX: Squadron/Signal, 2012.

Keane, John F., and C. Alan Easterling, CAPT USN. "Maritime Patrol Aviation: 90 Years of Continuing Innovation." *Johns Hopkins APL Technical Digest* 24, Issue 3, 2003.

Kempe, Frederick. *Berlin 1961.* New York: Penguin, 2011.

Kennett, Lee. *A History of Strategic Bombing.* New York: Scribner's, 1982.

La Fay, Howard. "DEW Line: Sentry of the Far North." *National Geographic Magazine* 114, Issue 1, July 1958.

LaFeber, Walter. *America, Russia, and the Cold War, 1945–1971.* New York: John Wiley & Sons, 1972.

Lasky, M., Richard D. Doolittle, B.D. Simmons, and S.G. Lemon. "Recent Progress in Towed Hydrophone Array Research." *IEEE Journal of Oceanic Engineering* 29, Issue 2, April 2004.

Launch Control Center Titan 2 ICBM Missile Complex Tucson Arizona U.S.A. Video.

LeMay, General Curtis E., with MacKinlay Kantor. *Mission with LeMay.* Garden City, NY: Doubleday, 1965.

Lemon, Stanley G. "Towed Array History, 1917–2003." *IEEE Journal of Oceanic Engineering* 29, Issue 2, April 2004.

"Lockheed Missiles & Space Company, The Early Years." *Lockheed Horizons,* Issue 12, June 1983.

Logan, Don. *The Boeing C-135 Series: Stratotanker, Stratolifter, and Other Variants.* Atglen, PA: Schiffer, 1998.

Manke, Robert C. *Overview of U.S. Navy Antisubmarine Warfare (ASW) Organization During the Cold War Era.* Newport, RI: Naval Undersea Warfare Center Division, August 2008.

McFarland, Stephen L. *Conquering the Night: Army Air Forces Night Fighters at War.* Air Force History and Museums Program, 1998.

The MG-10 System: The Aircraft and Weapon Control System of the F-102A (F-0582). Hughes Aircraft film.

Miller, David. *An Illustrated Guide to Modern Sub Hunters.* New York: Arco, 1984.

Minuteman Missile Silo Now a National Park—VOA Story. VOA News. Video.

Moeller, Colonel Stephen P. "Vigilant and Invincible." *ADA Magazine,* May–June 1995.

Moore, Captain John E., and Commander Richard Compton-Hall. *Submarine Warfare: Today and Tomorrow.* Bethesda, MD: Adler & Adler, 1987.

Morenus, Richard. *DEW Line: Distant Early Warning—The Miracle of America's First Line of Defense.* New York: Rand McNally, 1957.

"NATOPS Flight Manual Navy Model SP-2H Aircraft (NAVAIR 01–75EEB-1)." United States Navy Naval Air Systems Command, Washington, D.C., 1969.

Neufeld, Jacob. *The Development of Ballistic Missiles in the United States Air Force 1945–1960.* Office of Air Force History United States Air Force, Washington, D.C., 1990.

Nike-X: ABM "Army Air Defense Command" (ARAD-COM). U.S. Army, *The Big Picture,* TV, 1966.

Nike-Zeus: Pershing Missiles. U.S. Army, *The Big Picture,* TV, 1962.

NORAD Command Briefing. United States Air Force film.

NORAD (North American Aerospace Defense Command). USAF Air Defense Command film.

Nordeen, Lon O. *Air Warfare in the Missile Age.* Washington, D.C.: Smithsonian Institution Press, 2002.

Norden Bombsight: Conduct of a Mission. Army Air Force Training Command film.

Norden Bombsight: Operation. Army Air Force Training Command film.

Norden Bombsight: Principles. Army Air Force Training Command film.

Ogletree, W.A., H.W. Taylor, E.W. Veitch, and J. Wylen. "AN/FST-2 Radar-Processing Equipment for SAGE." Proceedings of the Eastern Computer Conference, May 6–8, 1958, Los Angeles, 156–160.

Operation Headstart: United States Air Force Film Report. United States Air Force film, 1959.

Operation Skywatch. William E. Stanfill. United States Air Force film.

Orion, Guardian of the Seas. Robert L. Mehnert. Lockheed California Company film.

Owen, David. *Anti-Submarine Warfare: An Illustrated History.* Barnsley, South Yorkshire, Great Britain: Seaforth Publishing, 2007.

"The P-3C Orion." *Lockheed Orion Service Digest,* Issue 25, August 1972.

"P-3 Orion Inertial Navigation System." *Lockheed Orion Service Digest,* Issue 12, December 1966.

"P-3C Update I." *Lockheed Orion Service Digest,* November 1976.

"P-3C Update II." *Lockheed Orion Service Digest,* Issue 35, December 1977.

"P-3C Update III." *Lockheed Orion Service Digest,* Issue 43, August 1985.

Panopalis, Terry. *Warpaint Series No. 64: Convair F-102 Delta Dagger.* Bletchey, Buckinghamshire, UK: Warpaint Books, 2007.

Partners for Peace. National Park Service—Minuteman Missile National Historic Site. Video.

"Pilot's Operating Manual for Airborne Radar AN/APS-6 Series for Night Fighters CO NAVAER 08–05–120." United States Navy Department, Washington, D.C., 1944.

Prados, John. *The Soviet Estimate: U.S. Intelligence Analysis & Russian Military Strength.* New York: Dial Press, 1982.

Price, Alfred. *Aggressors,* vol. 4. *Patrol Aircraft vs. Submarine.* Charlottesville, VA: Howell Press, 1991.

_____. *Aircraft versus Submarine.* Annapolis, MD: Naval Institute Press, 1973.

"Radar Bulletin No. 8-A (RADEIGHTA) Aircraft Control Manual," Chapters 12 and 13. Chief of Naval Operations, 1951.

Reade, David. *The Age of Orion: Lockheed P-3, An Illustrated History.* Atglen, PA: Schiffer Publishing, 1998.

Redmond, Kent C., and Thomas M. Smith. *From Whirl-wind to MITRE: The R&D Story of the SAGE Air Defense Computer.* Cambridge, MA: MIT Press, 2000.

Riccioni, Everest E. Colonel USAF Retired. "Strategic Bombing: Always a Myth." *Proceedings Magazine* 122/11/1,125, November 1996.

SAC Command Post. United States Air Force film, 1964.

"The SAGE/BOMARC Air Defense Weapons System—Fact Sheet." New York: International Business Machines Corporation, Military Products Division, 1958.

Schaffel, Kenneth. *The Emerging Shield: The Air Force and the Evolution of Continental Defense 1945–1960.* Washington, D.C.: Office of Air Force History United States Air Force, 1991.

Scrivner, Charles L. *Aircraft Number 82—TBM/TBF Avenger in Action.* Carrollton, Texas: Squadron/Signal, 1987.

Seconds for Survival. The Bell System, film, 1959.

Sherry, Michael S. *The Rise of American Airpower: The Creation of Armageddon.* New Haven, CT: Yale University Press, 1987.

The Shield of Freedom. United States Air Force, 1963 film.

Skywatchers. United States Air Force film.

Sontag, Sherry, and Christopher Drew, with Annette Lawrence Drew. *Blind Man's Bluff: The Untold Story of American Submarine Espionage.* New York: Public Affairs, 1998.

The Steel Ring. U.S. Army, unknown. The Big Picture TV.

Stefanick, Tom. *Strategic Antisubmarine Warfare and Naval Strategy.* Lexington, Massachusetts: Lexington Books, 1987.

Stone, Melvin L., and Gerald P. Banner. "Radars for the Detection and Tracking of Ballistic Missiles, Satellites, and Planets." *Lincoln Laboratory Journal* 12, Issue 2, 2000.

Story of the System Training Program (STP). USAF Air Defense Command, 1956 film.

Sullivan, Jim. *Aircraft Number 68—P2V Neptune In Action.* Carrollton, TX: Squadron/Signal, 1985.

Talmadge, Marion, and Iris Gilmore: *NORAD: The North American Air Defense Command.* New York: Dodd, Mead & Company, 1967.

Target for Today (T.F. I-3384). War Department film, 1944.

Technologies for Ocean Acoustic (Sound) Monitoring: The Secret Weapon of Undersea Surveillance. Video.

"TM 9–1400–250–10/2 Operators Manual: Overall System Description (Improved Nike-Hercules Air Defense Guided Missile Systems and Nike-Hercules Anti-Tactical Ballistic Missile (ATBM) Systems)." Headquarters, Department of the Army, Washington, D.C., December 1960.

To Catch a Shadow. Barry Shipman. Lockheed California Company film.

Tour a Cold War Era Atlas Missile Silo. Frakes Productions. Video.

Tracking the Threat. U.S. Navy film, 1982.

Tsipis, Kosta, and Randall Forsberg: *Tactical and Strategic Submarine Warfare.* Cambridge, Massachusetts: The MIT Press, 1974.

A 20-Year History of the Antiballistic Missile Program. U.S. Army Ballistic Missile Defense Systems Command, 1976 film.

UFO: Friend, Foe, or Fantasy. CBS Reports, 1966. TV.

"United States Air Force Statistical Digest 1947." Head-quarters U.S. Air Force, Washington, D.C., August 1948.

"United States Air Force Statistical Digest 1948." Head-quarters U.S. Air Force, Washington, D.C., September 1948.

"United States Air Force Statistical Digest Fiscal Year 1951." Headquarters U.S. Air Force, Washington, D.C., November 1952.

"United States Air Force Statistical Digest Fiscal Year 1952." Headquarters U.S. Air Force, Washington, D.C., n.d.

"United States Air Force Statistical Digest Fiscal Year 1953." Headquarters U.S. Air Force, Washington, D.C., n.d.

"United States Air Force Statistical Digest Fiscal Year 1954." Headquarters U.S. Air Force, Washington, D.C., n.d.

"United States Air Force Statistical Digest Fiscal Year 1955." Headquarters U.S. Air Force, Washington, D.C., n.d.

"United States Air Force Statistical Digest Fiscal Year 1956." Headquarters U.S. Air Force, Washington, D.C., n.d.

"United States Air Force Statistical Digest Fiscal Year 1957." Headquarters U.S. Air Force, Washington, D.C., n.d.

"United States Air Force Statistical Digest Fiscal Year 1958." Headquarters U.S. Air Force, Washington, D.C., n.d.

"United States Air Force Statistical Digest Fiscal Year 1959." Headquarters U.S. Air Force, Washington, D.C., n.d.

"United States Air Force Statistical Digest Fiscal Year 1960." Headquarters U.S. Air Force, Washington, D.C., n.d.

"United States Air Force Statistical Digest Fiscal Year 1961." Headquarters U.S. Air Force, Washington, D.C., n.d.

"United States Air Force Statistical Digest Fiscal Year 1962." Headquarters U.S. Air Force, Washington, D.C., n.d.

"United States Air Force Statistical Digest Fiscal Year 1963." Headquarters U.S. Air Force, Washington, D.C., n.d.

"United States Air Force Statistical Digest Fiscal Year 1964." Headquarters U.S. Air Force, Washington, D.C., July 1964.

"United States Air Force Statistical Digest Fiscal Year 1965." Headquarters U.S. Air Force, Washington, D.C., October 1965.

"United States Air Force Statistical Digest Fiscal Year 1966." Headquarters U.S. Air Force, Washington, D.C., n.d.

"United States Air Force Statistical Digest Fiscal Year 1967." Headquarters U.S. Air Force, Washington, D.C., n.d.

"United States Air Force Statistical Digest Fiscal Year 1968." Headquarters U.S. Air Force, Washington, D.C., September 1968.

"United States Air Force Statistical Digest Fiscal Year 1969." Headquarters U.S. Air Force, Washington, D.C., February 1970.

"United States Air Force Statistical Digest Fiscal Year 1970." Headquarters U.S. Air Force, Washington, D.C., February 1971.

"United States Air Force Statistical Digest Fiscal Year 1971." Headquarters U.S. Air Force, Washington, D.C., February 1972.

"United States Air Force Statistical Digest Fiscal Year 1972." Headquarters U.S. Air Force, Washington, D.C., September 1973.

"United States Air Force Statistical Digest Fiscal Year 1973." Headquarters U.S. Air Force, Washington, D.C., July 1974.

"United States Air Force Statistical Digest Fiscal Year 1974." Headquarters U.S. Air Force, Washington, D.C., April 1975.

"United States Air Force Statistical Digest Fiscal Year 1975." Headquarters U.S. Air Force, Washington, D.C., April 1976.

"United States Air Force Statistical Digest Fiscal Year 1976." Headquarters U.S. Air Force, Washington, D.C., April 1977.

"United States Air Force Statistical Digest Fiscal Year 1977." Headquarters U.S. Air Force, Washington, D.C., May 1978.

"United States Air Force Statistical Digest Fiscal Year 1978." Headquarters U.S. Air Force, Washington, D.C., May 1979.

"United States Air Force Statistical Digest Fiscal Year 1979." Headquarters U.S. Air Force, Washington, D.C., June 1980.

"United States Air Force Statistical Digest Fiscal Year 1980." Headquarters U.S. Air Force, Washington, D.C., March 1981.

"United States Air Force Statistical Digest Jan 1949—Jun 1950." Headquarters U.S. Air Force, Washington, D.C., April 1951.

"United States Fleet Anti-Submarine and Escort of Convoy Instructions, Part I: Surface Craft." United States Fleet Headquarters of the Commander in Chief, Navy Department, Washington, D.C., May 1945.

"United States Fleet Anti-Submarine and Escort of Convoy Instructions. Part II: Aircraft." United States Fleet Headquarters of the Commander in Chief, Navy Department, Washington, D.C., May 1945.

"United States Objectives and Programs for National Security." NSC 68, April 14, 1950.

"U.S. Objectives with Respect to the USSR to Counter Soviet Threats to U.S. Security." NSC 20/4, November 23, 1948.

"USAF Summary 1981." Headquarters U.S. Air Force, Washington, D.C., n.d.

"USAF Summary 1982." Headquarters U.S. Air Force, Washington, D.C., n.d.

"USAF Summary 1983." Headquarters U.S. Air Force, Washington, D.C., n.d.

Vance, P.R., L.G. Dooley, and C.E. Diss. "Operation of the SAGE Duplex Computers." Proceedings of the Eastern Computer Conference, May 6–8, 1958, Los Angeles, pp. 160–163.

Veale, Thomas F. Major USA, ed. *Guarding What You Value Most: North American Aerospace Defense Command Celebrating 50 Years.* Washington, D.C.: U.S. Government Printing Office, 2008.

War Games. John Badham. Metro-Goldwin-Mayer Studios, 1983 film.

Watts, Anthony J., ed. *Jane's Underwater Warfare Systems Third Edition 1991–1992.* Coulsdon, Surrey, UK: Jane's Information Group, 1991.

Werrell, Kenneth P. *Archie, FLAK, AAA, and SAM: A Short Operational History of Ground-Based Air Defense.* Maxwell Air Force Base, AL: Air University Press, 1988.

Whitman, Edward C. "SOSUS: The 'Secret Weapon' of Undersea Surveillance." *Undersea Warfare* 7, Issue 2, Winter 2005.

Wilson, Gordon A.A. *NORAD and the Soviet Nuclear Threat.* Stroud, Gloucestershire, UK: Amberley Publishing, 2011.

Winkler, David F. *Searching the Skies: The Legacy of the United States Cold War Defense Radar Program.* Langley AFB, Virginia: United States Air Force Air Combat Command, 1997.

Winokur, Robert S., and Craig E. Dorman. "Anti-Submarine Warfare and Naval Oceanography." *Oceanus Magazine* 33, Issue 4, Winter 1990/91.

Wood, Derek, with Derek Dempster. *The Narrow Margin: The Battle of Britain and the Rise of Air Power 1930–40.* New York: Paperback Library, 1969.

Y'Blood, William T. *Hunter Killer.* Annapolis, MD: Naval Institute Press, 1983.

Zabetakis, Stanley G., and John F. Peterson. "The Diyarbakir Radar." Unknown CIA publication 8, Fall 1964.

Index